Scrum 10

To Helen, Isabella and Amelia - *David*
For every learning experience I've had - *Jim*
To Arwenna, Nina, Mum and Dad - *Jiten*

Introduction

Over the years, we have delivered many different training sessions and talks - from introductory workshops to conference papers. We have also run one of London's most popular Agile-focused meetups (The London Agile Discussion Group) which now has well over one thousand members and has met nearly every fortnight since 2012.

Even though the topics vary across these sessions, we have noticed that the same questions keep cropping up.

Although the theoretical answers are available in the many books already published, as we work hands-on with teams every day, we know that real life doesn't always work out how the theory books tell us it should. So our replies are a mixture of theory along with reflections from our wide-ranging experiences, having worked for all sizes of companies (from start-ups to global corporations) and in industries ranging from fashion to government, automotive to mass retailing.

In the following pages you will find a selection of the questions that have been asked over the years, along with our combined answers. Interestingly, we didn't always agree on the answer, which led to some long debates and highlights that there is often no single, easy answer to what seems like a simple question.

We wanted this book to serve two purposes: First, it had to be a book that you could read from beginning to end without the feeling of jumping around all over the place, and second, it had to be easy to reference. With that in mind, we've broken the book down into six distinct sections. We start with some background to the main material, and then take you through the three key areas of Scrum - events, roles, and artefacts. We finish by introducing some related concepts. Each section has been given an appropriate icon, and these appear on every page ensuring that you can quickly find the section you are looking for.

Happy reading.

Background

These chapters take you through the story of how Agile and Scrum were created, and define the values and principles of both.

p8

Events

Events are the heartbeat of any Scrum system. These chapters will talk you through each event, what the desired outcomes might be and provides insights into some of the more complex concepts that surround them.

p20

Roles

Agile has people at its core, while teamwork is the essence of any successful undertaking. These chapters help you understand each of the Scrum roles, their responsibilities and their place in the team, while discussing some of the finer points of teamwork.

p44

Artefacts

Artefacts are the key to how work is formed within Scrum, and they double up as a great communication tool. These chapters take you through each artefact and bust the myths surrounding the role of documentation within Agile.

p62

Estimation

How Long will it take? How much will it cost? These chapters discuss the basics of planning poker, velocity and release planning and give clarity to some of the more nuanced parts of the role estimation plays within the Scrum framework.

p88

General

These chapters explain some of the other terms that get used frequently but are not directly part of Agile or Scrum. There are also a number of organisations and companies that you will come across frequently, and we've explained their role in the evolution of Agile with Scrum.

p114

Background

These chapters take you through the story of how Agile and Scrum were created, and define the values and principles of both.

What is the Agile Manifesto?

In 2001, a group of seventeen software specialists met in a ski resort in Utah, USA. They came from a variety of backgrounds and had no unifying approach, each preferring their own technique (for example, Extreme Programming, Scrum, Crystal, Feature-Driven Development or Pragmatic Programming).

Many referred to approaches as "lightweight" in reference to the lack of oppressive and cumbersome processes. However, the group felt that this term was detrimental and agreed on the formal term "Agile".

Despite their differences, the seventeen also agreed that a small number of underlying beliefs united all their approaches, such as working software being more valuable than extensive documentation. They produced a list of four core values and twelve principles, which they entitled the Manifesto for Agile Software Development (Beck et al., 2001) but is more commonly known as the Agile Manifesto. We will refer to it as the Agile Manifesto within this book.

You can view the Agile Manifesto, and become a signatory, by going to agilemanifesto.org.

What are the Agile Manifesto's 4 core values?

The Agile Manifesto (Beck et al., 2001) says:

We are uncovering better ways of developing software by doing it and helping others do it. Through this work we have come to value:

* Individuals and interactions over processes and tools
* Working software over comprehensive documentation
* Customer collaboration over contract negotiation
* Responding to change over following a plan

That is, while there is value in the items on the right, we value the items on the left more.

For example:

Talking to people to tell them that you have completed a feature... rather than relying on people being notified by an automated email from a management tool. Face-to-face communication is most effective and builds relationships.

The best measure of progress is whether something actually works in a live environment... not how thoroughly you have documented its internal workings (or your intentions).

Rather than defining requirements, or the end product, up front... work with your customer to define the direction as you go and agree the best way to achieve the goal. While it's okay to have a plan, things change (for example, market conditions)... don't continue with the original plan if it no longer makes sense. You want to build what is right now, rather than what was right when you started.

What are the 12 principles of the Agile Manifesto?

There are twelve principles behind the Agile Manifesto (Beck et al., 2001). Here they are with an example of why they might be beneficial:

"Our highest priority is to satisfy the customer through early and continuous delivery of valuable software."

"Deliver working software frequently, from a couple of weeks to a couple of months, with a preference for the shorter timescale."

Both of these principles allow product teams to get feedback, and to start to get a return on investment, as early as possible.

"Welcome changing requirements, even late in development. Agile processes harness change for the customer's competitive advantage."

If the world changes, our original plan may no longer be the best option. Therefore, we need to be able to change direction. By enabling this flexibility, our customers can get ahead of their rivals!

"Business people and developers must work together daily throughout the project."

We should not be defining the whole product up front; together with the business, we should create a vision of where we want to go, then work collaboratively with the customer to deliver what is best. This allows a speedy change of direction if necessary.

"The most efficient and effective method of conveying information to and within a development team is face-to-face conversation."

Don't communicate by sending emails when you can talk to someone in person; it is rarely as effective.

"Build projects around motivated individuals. Give them the environment and support they need, and trust them to get the job done."

"The best architectures, requirements, and designs emerge from self-organizing teams."

Giving the team autonomy is better than micro-management. Decisions are made for the team, by the team: they are self-organising. The team knows how best to achieve the vision.

"Working software is the primary measure of progress."

If you want to know if something is done, look at whether it is working in production. End users don't care how close to going live something is; if they can't use it, it has no business value – not even if it is 99 per cent finished.

"Agile processes promote sustainable development. The sponsors, developers, and users should be able to maintain a constant pace indefinitely."

Maintaining a steady output of small, releasable features is not only a more maintainable working approach, but also helps you forecast when features might go live. Erratic, pressurised working practices are not only unpleasant, they make planning very difficult.

"Continuous attention to technical excellence and good design enhances agility."

If you cut corners, it will come back to bite you later. For example, if you create a poor foundation of code, then all future work building on top of this code will take longer.

"Simplicity – the art of maximizing the amount of work not done – is essential."

Do what is requested to fulfil the goal, then get it live. This does not mean cutting corners, it means not over-working it (also known as 'gold-plating').

"At regular intervals, the team reflects on how to become more effective, then tunes and adjusts its behaviour accordingly."

We can all improve. Even if you think you are working really well, there is always room for more improvement. Just look at Formula 1 teams: Review... change... monitor... review... change... monitor... and so on.

What is Scrum?

Scrum is a framework that can be used to help apply the core values and principles of the Agile Manifesto (Beck et al., 2001), and is especially effective when people are working with complex adaptive systems (i.e. uncertain, unpredictable, changing environments).

Unlike many other project management approaches, the Scrum framework is 'lightweight'. So, while heavyweight approaches demand a lot of up front documentation, detailed specifications, planning, scheduling and reporting, Scrum minimises these. It encourages splitting up a product into small segments, releasing these segments to customers as they are completed, then using customer feedback on the completed items as a guide for future work. This means that Scrum relies on multiple cycles of planning, building and monitoring.

Although Scrum prescribes roles, events and artefacts (all of which are discussed in greater detail throughout the book), it is a simple framework that gives some structure, rather than a complete set of prescriptive rules. Scrum provides flexibility so teams can tailor it to their needs, and relies on the self-organisation of teams rather than micro-management. However, this freedom comes at a cost: teams have to think for themselves about how they should work. This book aims to help clarify some of the common questions that arise.

Although iterative, cyclical approaches can be accredited to many others, the Scrum we all know is largely due to the work of Ken Schwaber and Jeff Sutherland from the 1990s to date.

Why is it called Scrum?

The 'rugby approach' was first described by Hirotaka Takeuchi and Ikujiro Nonaka in a Harvard Business Review article in 1986 entitled The New New Product Development Game (Takeuchi and Nonaka, 1986). Having observed teams at companies such as Honda, 3M, Epson and Canon, they pulled together what they considered key attributes of successful teams working on new products. "Like a rugby team", they wrote, "the core project members at Honda stay intact from beginning to end and are responsible for combining all of the phases." We assume they are differentiating from American Football, which swaps over teams depending on the phase of play.

They explained "Companies are increasingly realising that the old, sequential approach to developing new products simply won't get the job done. Instead, companies in Japan and the United States are using a holistic method - as in rugby, the ball gets passed within the team as it moves as a unit up the field."

There was only one mention of Scrum in this article. It was the title of a paragraph ("Moving the Scrum Downfield") which outlined six characteristics for managing new product development. These characteristics were:

1. Built-in instability (for example, encouraging "trial and error by purposely keeping goals broad and by tolerating ambiguity")
2. Self-Organising project teams
3. Overlapping development phases
4. Multilearning (for example, learning as individuals, as teams, as an organisation. Also "accumulating experience in areas other than your own")
5. Subtle control (for example, giving the team autonomy, but avoiding chaos by acts such as selecting the right people for the project, creating an open environment, encouraging teams to listen to users, tolerating mistakes)
6. Organisational transfer of learning (for example, sharing wisdom and "converting project activities to standard practice") Takeuchi and Nonaka went on to write 'The Knowledge-Creating Company: How Japanese Companies Create the Dynamics of Innovation' (Nonaka and Takeuchi, 1995) but they just expanded on their rugby-like ideas rather than Scrum.

In 1995, Ken Schwaber delivered a paper at OOPSLA, in Texas, entitled 'SCRUM Development Process' (Schwaber, 1995). This references Takeuchi and Nonaka's earlier work, but is seen as the basis of Scrum as we know it today.

Is Scrum the same as Agile?

Agile is the underlying values and principles that Scrum lives by as documented in the Agile Manifesto (Beck et al., 2001).

You cannot apply the Scrum framework without the principles of Agile; you can use the meetings, roles and artefacts that Scrum describes but unless you apply the core principles of Agile at the same time, you will be delivering a product traditionally, just in smaller bursts.

However, you can be Agile without Scrum. There are number of other approaches that agree with and uphold the Agile Manifesto's values and principles, but don't use Scrum's roles, events, etc.

A well-known example of another Agile approach is Extreme Programming.

What are the three pillars of Scrum?

The three pillars of Scrum are transparency, inspection and adaption.

They are called 'pillars' because they are the foundation that uphold Scrum's empirical approach (that is, based on practice and experience).

Transparency
The team, and organisation as a whole, is working collaboratively towards a shared goal, so making the working practices open and explicit reduces misunderstandings. An example is setting a team Definition of Done, which is then shared and visualised so that everyone has a clear understanding of what it means for a piece of work to be done.

Inspection
The team is regularly reflecting on where it is in terms of how it is operating, encourages the team to maintain its focus on continuous improvement and helps it to achieve its goals better. As well as the team's end of Sprint retrospective, another example is reviewing the Sprint's progress daily, and then deciding as a team what needs to happen in order to get the work to done.

Adaption
Following on from inspection, this pillar encourages making changes when necessary and/or beneficial. An example could be making changes to the team's approach to testing if it is found that bugs were slipping through into production.

Inspection and adaption are fundamental focuses in the Scrum events of Sprint Planning, Daily Scrum, Sprint Review and Sprint Retrospective.

What are the Scrum values?

There are five Scrum values:

Commitment
The Scrum Guide™ (Schwaber and Sutherland, 2016a) requires that "People personally commit to achieving the goals of the Scrum Team."

This means that everyone in the Scrum team is committed to doing the very best they can to achieve the goals that the team has agreed to (based on what they knew at the time of commitment). However, it also means a commitment to improving themselves, the team and the organisation.

Focus
The Scrum Guide requires that "Everyone focuses on the work of the Sprint and the goals of the Scrum Team."

It's easy to get side-tracked. This value refers to staying focused to complete what you have committed to in each Sprint.

Openness
The Scrum Guide requires that "The Scrum Team and its stakeholders agree to be open about all the work and the challenges with performing the work."

To get where you need to go, everyone needs to have a voice and therefore you need transparency and openness from the start. But this value is also about communicating our progress honestly and clearly, and acknowledging any problems that we encounter on the way. Without openness, it is difficult to work as a team and even harder for us to improve.

Respect
The Scrum Guide requires that "Scrum Team members respect each other to be capable, independent people."

When we start a relationship on the understanding that our team members are able, but also accept that they have weaknesses, we will work better as a team.

Courage
The Scrum Guide requires that "The Scrum Team members have courage to do the right thing and work on tough problems."

None of the other values works without having courage; to stay focused and committed, to be open and respectful, all take strength and courage. This is not only applicable to the team, but to managers and the rest of the organisation.

The values were first mentioned in Ken Schwaber's 2004 book, Agile Project Management with Scrum (Schwaber, 2004) but were highlighted in the 2016 refresh of The Scrum Guide.

Schwaber has described the values as being like "the life blood of Scrum" (Schwaber and Sutherland, 2016b) because they are what makes Scrum alive. He went on to say that embracing the values helps people work together in teams: "With [the values], Scrum is a place you want to live; without them, it's not necessarily a place you want to be".

Events

Events are the heartbeat of any Scrum system. These chapters will talk you through each event, what the desired outcomes might be and provides insights into some of the more complex concepts that surround them.

What is the difference between a Sprint and an Iteration?

Sprints are a Scrum concept. The Scrum Guide™ (Schwaber and Sutherland, 2016a) says that Sprints run for one month or less. Our experience is that two-week Sprints are most common.

Iterations are an Extreme Programming (or XP) concept. XP recommends "iterations of 1 to 3 weeks in length".

Regardless of which word you use (we will use Sprint), they are both a planning cadence, which, due to their shortness, limits the risk of producing something that is no longer needed or fit for purpose. If your Sprint is two weeks long, the worst that can happen is that you spend two weeks building something that nobody wants; better than wasting months or years, which happens when you use some other approaches!

The team agrees what work it will undertake during the Sprint, with the aim of producing working software that can be put into production by the end of the Sprint. The Product Owner should have the option to release completed work at the end of each Sprint.

It doesn't matter which day the Sprint starts (it doesn't have to be a Monday) but Sprints should be regular and consistent in length. In other words, they are time-boxed, which means that the duration is set and does not change, regardless of the progress made (or not made) during the Sprint. Each successive Sprint should have the same duration.

Sprints contain all other meetings, including Sprint Planning, Daily Scrums, Sprint Reviews, Sprint Retrospectives, as well as the actual development work. A new Sprint begins immediately after the preceding Sprint finishes.

What is sprint zero?

Sprint zero is not an official Scrum event and is not mentioned in The Scrum Guide™ (Schwaber and Sutherland, 2016a). However, some teams use the event at the start of new projects.

It is a time-boxed period, usually the same length as future Sprints, in which the team can get the non-feature-based work done. These are, for example, setting up environments, agreeing the testing framework, building the Product Backlog. For this reason, Sprint zero might not include any potentially shippable features and may have a low velocity compared to future Sprints.

The aim is to provide a stable foundation for future work, which we think is a sensible thing to do - whether or not you try to complete it within a dedicated Sprint or do it over the first few Sprints.

Why are all events timeboxed in Scrum?

Setting a fixed, maximum duration to an activity is often referred to as 'timeboxing'. It is used in Scrum for all events (that is, Sprints, Sprint Planning, Daily Scrum, Sprint Review, Sprint Retrospective).

Once the event begins, the allocated time cannot be extended, but may be shortened if the goal is achieved early.

By keeping meetings short, timeboxing reduces waste. By limiting Sprints to between one and four weeks, timeboxing encourages more frequent delivery.

Sticking to the agreed timebox can be challenging, so it demands determination from the team and good facilitation skills from the Scrum Master.

How long should a Sprint be?

In the early days, Sprints were referred to in terms of months. The Scrum Guide™ (Schwaber and Sutherland, 2016a) now says that Sprints should have “a time-box of one month or less ...”.

But why are Sprints so short?

Change. The needs and desires of our customers, users, suppliers, stakeholders and Product Owners change rapidly and we need to take that into account. For example, what we were asked for six months ago may now be obsolete. What was relatively easy before, may have become more complicated and needs to be tackled in a different way. What was once safe, may be a much riskier undertaking now.

By regularly inspecting what we are doing, we can ensure that we change direction when necessary and beneficial. This adaptability often gives us competitive advantage by allowing teams to keep ahead of their rivals. Also, if everything goes wrong and what you've been working on ends up in the bin, it means that you've only wasted a few weeks!

This approach supports the Agile Manifesto (Beck et al., 2001) principle of delivering “frequently, from a couple of weeks to a couple of months, with a preference to the shorter timescale”.

Our experience is that the most common duration of a Sprint is two weeks. This allows the team to find their rhythm and release pretty regularly (although there's nothing to say you can't release continuously), without sacrificing the ongoing introspection and adaption that is fundamental to Agile.

What are the benefits of shorter Sprints? What are the benefits of longer Sprints?

Nobody can say whether 1, 2, 3 or 4 week Sprints are best for you as it depends on your situation. Your team needs to make that decision for itself, based on how frequently you want to reprioritise and plan, as well as your need for change.

Longer Sprints mean fewer meetings, but the meetings you do have are likely to be longer. They also mean that you have less ability to change direction and the risk that you are not working on the most valuable items is increased.

Shorter Sprints give you more flexibility, allowing you to change direction more often, and reduce risks of wasting time (because you will be reviewing your position more often). You will encounter more meetings, but hopefully they will be shorter than if you had longer Sprints.

Ask yourselves why you are thinking of changing the length of your Sprint. Is the duration of the Sprint really the problem or the best solution? Is there an underlying problem that you should be addressing instead?

For example, if your team is struggling to complete Product Backlog Items (PBI) in your current Sprints, then it might be better for the team to focus on breaking stories down and getting them to swarm on fewer PBIs at a time, rather than increasing your Sprint length. Equally, if you're finding that requirements change mid-Sprint, maybe consider whether you're accepting work without fully understanding the requirements, rather than making your Sprints shorter.

Finally, don't keep changing the length of your Sprints too often. Change sometimes takes a while to reveal the benefits. Wait a few Sprints before judging any change a success or a failure.

What is an increment?

At the end of each Sprint, the team should have a batch of freshly completed Product Backlog Items that are customer-ready.

A batch of completed PBIs, when combined with everything that has been completed previously makes up 'the increment'. In other words, each increment is cumulative because it comprises the most recent Sprint's output with the entire previous Sprints' output.

Each increment must be 'potentially shippable', meaning that the new batch of PBIs plus all previously completed PBIs should work together and be in a state where the new work can be released to customers if the Product Owner chooses. If the Product Owner chooses not to release at the end of a Sprint, then the next increment will include all these unreleased items and the next Sprint's completed items.

The Scrum team as a whole plans what will be in each new increment at the start of the Sprint, and the Development Team works together to turn the Sprint Backlog Items into an increment during the Sprint.

Do I have to release code to live at the end of each Sprint?

No.

However, work should be 'done' at the end of each Sprint. This means that each PBI can be released to live (also called production), if desired by the Product Owner. As The Scrum Guide™ (Schwaber and Sutherland, 2016a) says, each completed PBI "must be in useable condition regardless of whether the Product Owner decides to actually release it." The common phrase that people use is 'potentially shippable increment'.

The Product Owner and Development Team should agree on the policy of releases. This will form the basis of the release plan, which will outline when it is hoped that features will be released.

The Agile Manifesto (Beck et al., 2001) encourages frequent, early and continuous delivery because the sooner you release work, the sooner you get feedback on what you've built. Holding items back from being released isn't making the most of working with Scrum, so we'd encourage you to ask why code can't be released at least at the end of each Sprint.

Can a Sprint be cancelled?

Although anyone can suggest a Sprint should be cancelled, only the Product Owner can actually cancel it. This can be done at any time during the Sprint.

One of the main reasons that Sprints are short is to provide flexibility on what is being developed. Product Owners should not be cancelling a Sprint just because they want the team to do something different. Cancelling a Sprint should be considered a last resort.

However, if the Sprint Goal and Product Backlog Items in the Sprint Backlog are no longer relevant (for example, customers no longer need the features that you are building), then aborting the Sprint makes sense. But, if you are working in a very turbulent market and cancelling Sprints happens often, you may want to consider shortening your Sprints to maximise the flexibility available rather than cancelling Sprints and wasting a lot of time and effort spent planning and re-planning.

If a Sprint is cancelled, all unfinished work is returned to the Product Backlog. The new Sprint is then planned as usual. Any PBIs that were completed before the Sprint was cancelled may be released if the Product Owner believes they still provide value.

What is Sprint Planning?

Sprint Planning is a meeting in which the whole team plans what it will be working on over the next Sprint.

The session is time-boxed: The Scrum Guide™ (Schwaber and Sutherland, 2016a) recommends limiting it to "a maximum of 8 hours for a one-month Sprint".

The meeting starts with the Product Owner explaining what s/he wants to achieve in the next Sprint and identifying which Product Backlog Items in the Product Backlog will help achieve this. If they've not already done so, the team will work with the PO to understand the PBIs, estimate them and break them down into achievable chunks. The team then agrees what they will undertake in the Sprint, using velocity as a guide.

By the end of the Sprint Planning session, the team will have agreed which PBIs they will tackle in the Sprint and how they intend to get these released to a production environment. This forms the Sprint Backlog. The team will also set a Sprint Goal.

Is Sprint Planning the same as release planning?

No.

Sprint Planning is short-term and focused on the next few weeks' work.

Ideally, you will be releasing each work item as soon as it has been completed. That way, you start reaping the benefits straight away.

A release plan is a high-level, longer-term view of the project, which is much less precise: it cuts the project into manageable parts and maps out the order of these chunks using very rough estimates.

Having such a plan can help direct the team. For example, when prioritising the Product Backlog, creating Sprint Goals and building Sprint Backlogs.

However, please don't think of release plans (or any other plans) as fixed or immutable; our approach welcomes change and gives us flexibility when advantageous.

What is the point of the Daily Scrum?

The purpose of the Daily Scrum (also known as Daily Stand-Up, Stand-Up or just The Scrum) is to help the team work together to achieve the Sprint Goal and complete the work in the Sprint Backlog.

It is held daily and focuses on recent progress, the current position and immediate future: specifically what has happened since the Daily Scrum and what is expected to be done before the subsequent one.

The Scrum Guide™ (Schwaber and Sutherland, 2016a) suggests the following format:

- What did I do yesterday that helped the Development Team meet the Sprint Goal?
- What will I do today to help the Development Team meet the Sprint Goal?
- Do I see any impediment that prevents me or the Development Team from meeting the Sprint Goal?

It is held at the same time and place each day, ideally close to the team's Scrum board. It is time-boxed to 15 minutes (any in-depth discussions should take place after the Daily Scrum).

The Scrum Master should ensure that the team has the meeting but everybody in the team should contribute. Although The Scrum Guide says "only Development Team members participate in the Daily Scrum", we feel that others can contribute if they will add value to the team's understanding of the work they are doing. However, in such instances, the Scrum Master is responsible for making sure people outside the team don't hijack the meeting!

What is a blocker?

A blocker can be anything that is stopping the team getting work done.

Blockers can be internal or external to the team. Although they are often caused by dependencies, this is not always the case.

An example of a blocker caused by an internal dependency would be waiting for another member of the team to change a database structure, supply some assets, run tests or release the code to the Production environment.

An example of a blocker caused by an external dependency would be waiting for another team or third party supplier to provide an API.

Although most blockers are caused by such dependencies, there are other types of blockers. For example, you might be blocked because you need an event to happen before you can complete the item, such as Christmas.

Many blockers are foreseeable. Unless the team agrees that the blocker can be resolved during the Sprint, we recommend not taking Product Backlog Items into the Sprint until the blocker has been removed.

Although the Scrum Master can help remove such blockers, it is best if the team can do it for themselves — otherwise you are just building in more dependencies within your team.

Visualising blockers helps teams keep them front of mind and focuses the team on removing them.

Can a Daily Scrum be longer than 15 minutes?

It should take as long as needed. No more and no less.

However, keeping it less than 15 minutes is recommended. If the team is under ten people, then this should not be too difficult.

If your team is under ten people, but you find yourself regularly exceeding 15 minutes, look out for:

- People going into too much detail
- Discussions that only one or two people are interested in
- People trying to solve problems
- Side conversations
- Lack of focus and/or structure of the stand-up
- Discussions that aren't relevant to the Sprint and/or team

It is easy for stand-ups to lose their focus. Remember, this is a chance for the team to remind themselves of their current focus. It might help if someone 'drives' the stand-up by looking out for the above signs and encouraging the team to question whether certain discussions could be held later. This is often done by the Scrum Master, but doesn't have to be; it could be a different person each day.

You might want to consider an alternative format: discuss each PBI that is in progress, rather than focusing on what each individual is working on. Start with the PBI that is nearest to completion and ask "What needs to be done to move this forwards?" After this has been discussed, move on to the next ticket nearest completion and repeat the process until all tickets in progress have been discussed.

Can you do a stand-up sitting down?

Where possible, you should stand up. Not only does it signify that the meeting is in effect (preventing accidental interruptions), it also encourages the team to come together rather than shout from their desks and brings a certain amount of energy to the discussion. Equally importantly, it encourages brevity; sitting down tends to prolong the meeting unnecessarily.

What is a Sprint Review?

This is a meeting to show people what has been completed during the Sprint and to get feedback. It doesn't need to be a polished presentation and is definitely not a training session on how to use what has been developed. Stakeholders and anyone else that might be affected and/or interested are invited.

The team and/or Product Owner explains which items from the Sprint Backlog have (and have not) been completed, and confirm what remains in the Product Backlog. After this, the room discusses what they think should be done next. This is useful for two reasons: it is an opportunity to change what is in the Product Backlog (for example, removing PBIs that are no longer relevant, adding new items, changing priorities), and it is valuable input for the next Sprint's planning session.

Each team runs their Sprint Review differently. Personally, we prefer the Development Team to show what has been completed, rather than the Product Owner, as it helps nurture collaboration between the team and the stakeholders. Also, the team is usually best placed to answer questions from the floor.

When should we hold a Sprint Review?

You should hold a Sprint Review at the end of each Sprint.

Sprint Reviews are a great opportunity to get feedback on what you have done during the last Sprint, to give the team's morale a boost by seeing the benefits their work has generated, and to receive input for what should be in the next Sprint Backlog. Holding Sprint Reviews less frequently than every Sprint is a missed opportunity for useful, collaborative feedback.

There is nothing stopping you having Sprint Reviews more often, but it generally makes sense to have them at the end of each Sprint.

However, we are not advocating that you wait until the Sprint Review to get any feedback on what you are creating; talking to people, and showing them what you have created, throughout the Sprint is usually beneficial - especially the Product Owner. However, save the discussions about future Sprints for the Sprint Review.

Can I show unfinished items in a Sprint Review?

Not according to The Scrum Guide™ (Schwaber and Sutherland, 2016a) which says, "The Development Team demonstrates the work that it has 'Done'...".

However, most people agree that, if you have something relevant and want to get feedback from certain stakeholders who are otherwise difficult to contact, you might want to present it in order to get input.

But please be careful if you are showing unfinished work: make sure you are very clear about what stage it is at and what is left to do before it is complete; people often hear what they want to hear, and will consider work shown in a Sprint Review as finished.

If in doubt, stick to presenting work that is 'done'.

What is a Sprint Retrospective?

The twelfth principle of the Agile Manifesto (Beck et al., 2001) is: "At regular intervals, the team reflects on how to become more effective, then tunes and adjusts its behavior accordingly."

Sprint Retrospectives (aka Retros) offer this opportunity to inspect and adapt. Each Retrospective is time-boxed and should be held at the end of each Sprint, before the next Sprint's planning session.

The whole team should be involved, including the Product Owner and Scrum Master. The Product Owner is present only in the capacity as one of the team, not as a dominant force in the meeting. The Scrum Master usually plays the role of facilitator.

A Retrospective's focus is to consider how well the newly concluded Sprint went. In terms of processes, collaboration, tools, etc, not in terms of judging the results of what was produced (keep that for the Sprint Review) or an individual's performance. There are many ways of running Retrospectives but the broad goals, regardless of style, are to identify:

- What went well (so you can keep doing it), and
- Where there is room for improvement.

The team decides what it wants to change in order to improve. We recommend the output of all Retrospectives is at least a number of action points with timescales and name(s) of who will be working on it, against each action. These are not improvements for the Scrum Master to implement; if the team wants something to change, they will actually have to change something.

Teams should not necessarily wait for the Retrospective to make improvements. However, we recommend that significant changes should be discussed as a team before being implemented, and the Retrospective offers a great opportunity without disturbing the team too frequently.

What is the Prime Directive?

"Regardless of what we discover, we understand and truly believe that everyone did the best job they could, given what they knew at the time, their skills and abilities, the resources available, and the situation at hand." – The Prime Directive

Originating from Norman L. Kerth's book, Project Retrospectives: A Handbook for Team Reviews (Kerth, 2001), it is used as a foundation for Sprint Retrospectives. The Prime Directive is meant to instil a positive mindset, reassuring the group that the retrospective will be conducted openly, with trust, non-judgmentally and without apportioning blame.

Kerth tells a story (Rising, 2008) of how every aspect of a sailing race was reviewed after a competitor died. Nobody was looking for a scapegoat or to find fault or judge others' actions; they were looking for lessons so they could prevent future fatalities in a race. He wanted to see more of that type of introspection amongst teams.

Facilitators will often read the Prime Directive out loud at the start of the Sprint Retrospective, and then put it up on the wall so that it's visible throughout. However, our experience is that you cannot rely solely on the Prime Directive to instil feelings of trust, openness, collaboration, etc.; you should look for other methods of encouragement and use them in conjunction with the Prime Directive.

Although the Prime Directive is mainly used for Sprint Retrospectives, there is no reason to limit its use to only those events.

Why are Scrum meetings called ceremonies?

The Oxford Dictionaries' (Oxford Dictionaries I English, 2016a) definition of a ceremony is: "An act or series of acts performed according to a traditional or prescribed form", but the ceremonies described in Scrum are far from traditional. While Sprint Planning, the Daily Scrum and the Sprint Retrospective can all be run with a similar format each time, in order to improve and to simply keep the sessions from going stale, we suggest that you mix things up a bit. Scrum allows the team to decide many of the details, filling in the blanks that the framework itself does not prescribe.

Certain details are stated in The Scrum Guide™ (Schwaber and Sutherland, 2016a) , such as the Daily Scrum being "a 15-minute time-boxed event", and it favours the term 'events' over 'ceremonies'. It states: "Prescribed events are used in Scrum to create regularity and to minimise the need for meetings not defined in Scrum".

'Events' does seem to be a better way of describing the cadence of Scrum, and feels less likely to put off those already sensitive to its prescribed nature.

What is a Sprint Goal?

A Sprint Goal helps the team know what it is aiming to achieve by the end of the Sprint, thus offering guidance to the team when working on PBIs. For example, if wondering whether to do A or B, the team should ask the question: "Which of the options is most likely to achieve the Sprint Goal?", then take that route.

The Sprint Goal is created by the team along with the Product Owner. It is based upon the PBIs in the Sprint Backlog. Once the PBIs in the Sprint Backlog are completed, the Sprint Goal should also have been reached.

As The Scrum Guide™ (Schwaber and Sutherland, 2016a) says, the Sprint Backlog can change during the Sprint as the team "learns more about the work needed to achieve the Sprint Goal". If the team considers a PBI unnecessary to achieve the Sprint Goal, they should discuss with the Product Owner as to whether the item should be removed from the Sprint Backlog. However, as The Scrum Guide™ goes on to say, "No changes are made that would endanger the Sprint Goal."

If the Sprint Goal becomes invalid or unnecessary, then the Product Owner may cancel the Sprint entirely.

How do you plan a Sprint when you have different skill sets in the team?

Different skills are usual within a team. The term used in Scrum is “Cross-functional”. A team is considered cross-functional, as written in The Scrum Guide™ (Schwaber and Sutherland, 2016a), when it has “all competencies needed to accomplish the work without depending on others not part of the team”. That does not mean that each member of the team holds all the skills.

Sprints are planned based on the highest priority work, guided by velocity. The team as a whole works towards completing the Sprint Goal.

However, teams should try to learn skills from each other as they mature. This reduces the dependency on one or two individuals to perform certain tasks. A popular approach to share information and skills is pair programming.

Roles

Agile has people at its core, while teamwork is the essence of any successful undertaking. These chapters help you understand each of the Scrum roles, their responsibilities and their place in the team, while discussing some of the finer points of teamwork.

How many people should there be in a scrum team?

The Scrum Guide™ (Schwaber and Sutherland, 2016a) states, "Optimal Development Team size is small enough to remain nimble and large enough to complete significant work within a Sprint. Fewer than three Development Team members decrease interaction and results in smaller productivity gains".

Many people in the Agile community have adopted the idea that seven plus or minus two people (i.e. five to nine people) is the optimal size for a scrum team. The idea has its origins in psychology and, more specifically, is taken from a paper written by George A. Miller called The Magical Number Seven, Plus or Minus Two: Some Limits on Our Capacity for Processing Information (Miller, 1956).

In general, software is most effectively developed by small crossfunctional teams: forming teams that contain everyone required to deliver software autonomously is optimal. Combining the two ideas above, that would require somewhere between three and nine people.

The drawbacks of a team that is too small appear obvious, but a team that is too large can also create problems. Sprint Backlogs become bigger and more difficult to manage, leading to a lack of clarity on progress. Daily Scrums, Sprint Retrospectives and planning sessions will all take longer, and those ceremonies will become more difficult to facilitate.

Whatever the size of the development team, the Scrum team will only contain one Product Owner and one Scrum Master.

What is the difference between the 'Development Team' and the 'Scrum Team'?

The Development Team includes developers, manual and automation testers, quality assurance, user experience, designers, et al. We don't differentiate between people doing any of these functions in terms of the three roles in Scrum; they are all grouped together as the 'Development Team'.

The Scrum Team includes everyone in the Development Team plus the Product Owner and Scrum Master.

Why is there no tester in a Development team in Scrum?

It's not that there isn't a tester, it's that The Scrum Guide™ (Schwaber and Sutherland, 2016a) doesn't distinguish between specialisms. It recognises that testing as an activity exists, but wants to promote team accountability.

The Scrum Guide states: "Development Teams have the following characteristics:

- They are self-organising. No one (not even the Scrum Master) tells the Development Team how to turn Product Backlog into Increments of potentially releasable functionality;
- Development Teams are cross-functional, with all of the skills as a team necessary to create a product Increment;
- Scrum recognises no titles for Development Team members other than Developer, regardless of the work being performed by the person; there are no exceptions to this rule;
- Scrum recognises no sub-teams in the Development Team, regardless of particular domains that need to be addressed, such as testing or business analysis; there are no exceptions to this rule; and,
- Individual Development Team members may have specialised skills and areas of focus, but accountability belongs to the development Team as a whole."

Every Development Team is different and it is up to the team to decide who they need on it, and how activities such as testing are approached.

What is the hierarchy in a Development Team?

There is no hierarchy within the Development Team; the team as a whole makes decisions about how they will work (for example, what work they will take into the Sprint Backlog). The team succeeds or fails as a unit.

Much research has shown that groups often work better without a line manager being in the team. However, this doesn't mean that they need to work in isolation from other teams or that different levels of knowledge can't be held within the team.

Although the Product Owner and Scrum Master are technically not members of the Development Team, there is still no hierarchy there either: the Product Owner helps guide what the team will work on, but the Development Team decides how they will complete it; the Scrum Master is there to help the Development Team achieve their goals.

What is a Scrum Master?

The Scrum Master is responsible for making sure the Scrum Team understands and applies the Scrum values and the principles of Agile.

A Scrum Master can assist a Scrum Team in a number of different ways. They can assist a Product Owner by helping them to understand how to create clear PBIs, and with the management, prioritisation and review of the Product Backlog on an ongoing basis. They can help the Development Team by highlighting when the team might be undertaking harmful activities (for example, overcommitting with a Sprint backlog), and can coach them to work effectively, efficiently and consistently through the use of Scrum. They can remove impediments that arise, or help the Development Team to remove them for themselves, and they can encourage the team to improve itself towards self-organisation and a higher level of achievement. The Scrum Master is also a facilitator, helping to ensure the Scrum Ceremonies are effective, and by making sure they actually take place!

An effective Scrum Master will also have a big part to play in helping organisations adopt Scrum. They are capable of breaking down barriers to adoption not just within software delivery environments, but also in the wider organisation, ensuring that ways of working are transparent, understood and that the interactions between teams are productive.

Is the Scrum Master or Product Owner senior to the development team?

Simply put, no. The Scrum Master, the Product Owner and the Development Team all sit at the same level, they are all part of the same team, and they all work for each other.

Scrum Masters are often referred to as the team's 'Servant Leader'. The term was introduced by Robert K. Greenleaf in his 1970 essay entitled The Servant as Leader (Greenleaf, 1970), and we use it to mean that, unlike a military general who commands troops, Scrum Masters are more like a revolutionary leader: requested by the team to help them towards a better future. Change is by the team, for the team. Scrum Masters may work at various levels. They can help both individuals and teams with self-organisation, autonomy and purpose, and they can help departments or corporations adopt better practices across their organisation.

Product Owners are responsible for ensuring that the team is always delivering maximum value and essentially define the 'what' and the 'why'. The PO needs feedback from the Development Team in order to help them do their job effectively, and the Development Team needs the PO to provide clear direction for the product being built. They work together with the Development Team as members of the Scrum Team.

What is the difference between a Scrum Master and a project manager?

This depends on your definition of a project manager. The UK Government (nationalcareersservice.direct.gov.uk, n.d.) describes a project manager's responsibilities as:

- Finding out what the client or company want to achieve
- Agreeing the timescales, costs and resources needed
- Selecting and leading a project team
- Drawing up a detailed plan and schedule for each stage of the project
- Plan and manage... projects, and make sure they are completed on time, meet the needs of the client and stay within budget.

The project manager described above is very much managing the team: s/he is making agreements on the team's behalf and will then push the team to deliver on that promise... on time and in budget!

Scrum Masters are not the boss of the team and they do not commit on the team's behalf; they work with the team to complete the project.

However, just because a project is being run in an Agile way using Scrum, it doesn't mean that questions about planning, scheduling, costing, budgeting, etc, disappear. However, unlike the stereotyped project manager, the Scrum team as a whole will cover these topics. For example, the Product Owner is responsible for establishing what clients want; the Development team is responsible for quality and agreeing to timescales.

The Scrum Master is responsible for making sure that the Scrum Team addresses such questions.

Should the Scrum Master have technical knowledge?

Some people think that a Scrum Master should be able to coach the team in technical matters; others say that the team is responsible for improving technical skill and that a Scrum Master with such knowledge risks getting lost in the minutiae.

Some people say that the Scrum Master's role is to help the team get better; others say that helping the team improve might require technical knowledge.

Truth is, every team is different and has different demands on its Scrum Master.

Our opinion is that it is more important for a Scrum Master to have 'people skills' than technical knowledge and that it is rare for a Scrum Master to need technical knowledge in order to:

- Understand enough about the goals and work that the team is doing
- Help the team get impediments removed (Note: This doesn't mean they can remove the impediment, just that they can speak to the right people to help resolve the issue)
- Lead the team in continuous improvement.

Is it the Scrum Master's role to remove impediments?

Although it's a subtle difference, we'd say that it is the Scrum Master's responsibility to help the team remove impediments themselves. It's not about the team throwing problems at the Scrum Master and expecting them to be solved.
The Scrum Master's role is generally two-fold.

Firstly, it is to help the Scrum team remove barriers that stop the team from running efficiently. Examples include stopping a constant stream of questions coming from outside the team (often referred to as 'noise reduction'), making sure that the team has the right tools, environment and support needed to achieve their goals, helping the team resolve issues regarding these areas, helping connect people together when needed (often known as a 'network broker').

Secondly, the Scrum Master helps the team 'grease the wheels' so it functions more effectively. Examples include coaching the team towards further improvements through regular inward reflection, training, being an 'Agile evangelist' to extend the use of Agile across the organisation.

What is a Product Owner?

The Product Owner (or PO) is the person who understands the business and customers, and has the authority to make decisions about the project. The PO is responsible for knowing 'why' the project is being undertaken and 'what' is needed to achieve this goal. The PO might get this information from internal users and/or external customers; they are not expected to know it off the top of their head!

But the PO is responsible for listening to these demands, deciding what is needed, and relaying that to the rest of the Scrum team. The PO is accountable for building a list of requirements that the team will work on (this list is known as a 'Product Backlog' and the work items in it are known as 'Product Backlog Items' or 'PBIs'). They are also accountable for prioritising the PBIs within the Product Backlog, usually based on the value that the PO believes will result from the item being completed.

In short, the PO is ultimately responsible and accountable for deciding what is to be built and in what order.

Does the Scrum Master plan work with the Product Owner?

No, the Scrum Master is not responsible for what is worked on or when. The Product Owner defines what is worked on and when, and the Development Team defines how it will be built and how long that will take.

In reality though, the Product Owner may well approach the Scrum Master and the Development Team for their opinions. The Scrum team should work as a unit, collaborating to ensure that the right thing is delivered, in the right way.

What is a Product Manager and where do they fit within Scrum?

Product Manager is not a role within Scrum. Therefore, you will need to clarify what 'you' mean by Product Manager to answer this question. Wikipedia's (Wikipedia, n.d.) description is: "A product manager communicates product vision from the highest levels of executive leadership to development and implementation teams. The product manager is often called the product CEO."

This sounds like the Product Manager could be someone who just communicates requirements, passing them on from the users, customers and company executives to the Development Team. The role of Product Owner in Scrum exceeds this description; POs listen to such needs, evaluate and prioritise them (with the help of the Development Team), acting as collaborator through the development process. Scrum's PO has the authority to make decisions around what and why such items are worked on.

It is best to think of a Product Owner as a role, not a job title. As long as the PO has knowledge of what is needed, and authority to make decisions about what is delivered and when, then it doesn't matter what their official job title is; they will still be the Product Owner for the team.

If you have multiple stakeholders, can you have more than one PO?

No. There will only ever be one Product Owner (PO) per team regardless of how many stakeholders there are.

The PO will gather stakeholder needs and present them as a single voice to the team in the form of unambiguous requirements. The PO also works with stakeholders to prioritise work against all other requests, ensuring that each stakeholder has a voice in those discussions, especially where different stakeholders priorities are competing.

What technical (development) knowledge does a PO need to have?

It is rare that a Product Owner needs to have technical knowledge; it is more important to have knowledge of customer needs and the authority to make decisions. Exceptions might include a situation where the product being built is for the software community (such as building GitHub), in which case technical knowledge might be required to understand the customer needs.

Although there are benefits of a Product Owner with technical knowledge, there are also benefits of a Product Owner without such knowledge.

A Product Owner with technical knowledge might:

- Relate better to the Development Team
- Understand the details of problems and/or limitations the Development Team is facing
- Provide useful technical ideas
- Better understand why development is taking longer than desired
- Better understand how the bigger picture fits together

A Product Owner without technical knowledge might:

- Be less likely to try to define the solution
- Be more likely to focus on WHY something is being built and for WHO, rather than look for the 'what' and 'how'
- Be more likely to push for an early, usable product rather than a perfect technical solution

These are generalisations and are only used to illustrate that there are benefits found in all sorts of backgrounds.

What is a cross-functional team?

'Cross-functional' is a term that is often misunderstood and misused. It does not mean that every member of the team can do 'everything' needed to complete work in the Sprint Backlog.

It actually means that the team should contain individuals with sufficient skills so that, as a group, it is able to complete the Sprint Backlog without the help of people outside the team. Therefore, for example, the team will not rely on someone outside the team to release code to production.

However, it certainly doesn't intend to encourage individuals to have a blinkered focus only on their area of expertise; sharing knowledge will make the team stronger and more productive.

Artefacts

Artefacts are the key to how work is formed within Scrum, and they double up as a great communication tool. These chapters take you through each artefact and bust the myths surrounding the role of documentation within Agile.

What are the artefacts of Scrum?

Scrum has 3 artefacts:

- Product Backlog
- Sprint Backlog
- Increment

Product Backlog
This is a collection of all the requirements that the Product Owner wants the product to contain. In effect, it is a prioritised list of everything still to do. There is only one Product Backlog per product.

Sprint Backlog
This is a sub-section of the Product Backlog, consisting of Product Backlog Items that the team will work on in a Sprint. It is created by the team picking a number of PBIs from the top of the Product Backlog. A PBI should only be taken into the Sprint Backlog if the team believes it can be completed before the end of the Sprint. There is an individual Sprint Backlog produced for each Sprint.

Increment
The increment is all of the PBIs that have been completed within the Sprint plus all the other PBIs previously completed (including the existing code in production). Everything within the increment should be considered 'Done' by the team, be production-ready and should work with the existing code base. However, it is up to the Product Owner as to whether s/he wishes to release the increment to production.

What is a Product Backlog?

A 'Product Backlog' is a list of everything that might be desired for the product at some point in the future. It should include new functionality but also improvements and fixes.

The Product Owner is responsible for creating and prioritising the Product Backlog. The PO is also responsible for updating and reprioritising the Product Backlog as items become obsolete and new items are discovered. While only the PO has this right to change it, anyone in the organisation with an interest should be able to view the Product Backlog. The PO can update the Product Backlog at any time.

Ideally, the Product Backlog should be a relatively small list of PBIs so that it is manageable. Requirements tend to get stale as time goes by, so regularly trimming the Product Backlog down to contain only things relevant at that time could be considered good practice for the Product Owner.

Higher priority PBIs often have more detail than lower priority items. The reason for this is that higher priority items will be worked on sooner, while lower priority items might not be worked on for weeks, months or years (by which time the requirements might have changed) or may never be needed. Therefore, it is common for lower priority PBIs to be left as large items, or 'epics', which will be broken down into smaller items when they get closer to being worked on. Once broken down, each of these fragments is classified as its own PBI.

Can multiple teams share the same Product Backlog or should they always have their own?

This is perhaps not a question of whether they should have their own Product Backlog, but more whether they could share one. The answer is undoubtedly context driven, so perhaps there are other questions that need to be answered first.

If there are multiple teams building the same product, then the prioritisation should be consistent across all of those teams, and a single Product Backlog could be possible.

Be careful to keep PBIs independent of each other so that the teams aren't treading on each others toes. This can be especially difficult when teams are set up to work on components (for example, a server-side team and a client-side team), rather than a multidisciplinary, product-focused team.

You want the teams to feel a sense of ownership for the product, or their part of the product, and the backlog or backlogs should be set up in a way that supports that.

What is a Product Backlog Item?

(Often abbreviated to PBI)

'Product Backlog Item' is the generic label given to a user story, epic, defect - in fact, anything that describes a change or addition that the Product Owner would like made to the product at some point in the future.

Each PBI's description is short and simple, giving enough information for the team to start a conversation about what is needed. This brevity encourages collaboration between the Development Team and Product Owner/stakeholders to find a suitable solution. PBIs are written by the Product Owner - with help from the team - using clear, everyday language that everybody can understand.

Higher priority PBIs often have more detail than lower priority items. The reason for this is that higher priority items will be worked on sooner, while lower priority items might not be worked on for weeks, months or years (by which time the requirements might have changed) or may never be needed. Therefore, it is common for lower priority PBIs to be left as large items, or 'epics', which will be broken down into smaller items when they get closer to being worked on. Once broken down, each of these fragments is classified as its own PBI.

Is a detailed spec still written?

No. Scrum doesn't have extensive up-front documentation that heavyweight methodologies have (for example, PRINCE2's Project Initiation Documentation).

Instead, the Scrum framework uses Product Backlog Items. These provide enough information to get the team started but their limited detail encourages collaboration with stakeholders and the Product Owner, the latter confirming the aims and goals, while the team defines how the aims and goals will be achieved. Hence, why PBIs are often referred to as a mere starting point for a conversation.

More detail is often added to a PBI as it gets closer to being worked on; there's no point in giving anything more than a very high-level description for items towards the bottom of the queue as the aims and goals may change by time you get to them or they may never be worked on at all.

This approach has a number of benefits including: transparency, flexibility (the PBI can be updated as more is learned about the product), better outcome (by focusing on the goal rather than the solution, and the team is not wedded to a pre-determined approach).

What is a Sprint Backlog?

The 'Product' Backlog lists all outstanding items that the Product Owner would like to be completed before the end of the project. During Sprint Planning, the team forecasts that they will complete a number of these Product Backlog Items during the Sprint: this collection of selected PBIs (along with a plan for getting them into production) is what is known as the 'Sprint Backlog'.

It is important to note that the team agrees on the items that go into the Sprint Backlog; it is not just a selection made by the Product Owner. However, the items chosen will predominantly be the highest priority items from the Product Backlog, although it's created with the Product Owner to ensure that the PBIs in the Sprint Backlog are relevant to, and focused around, the Sprint Goal.

The team should work on each PBI in the Sprint Backlog roughly in priority order. However, the team is self-organising, so they know how best to approach each Sprint.

Can you add a Product Backlog Item to the Sprint Backlog after the Sprint starts?

The Scrum Guide™ (Schwaber and Sutherland, 2016a) states: "The Sprint Backlog is the set of Product Backlog items selected for the Sprint, plus a plan for delivering the product Increment and realizing the Sprint Goal."

The Development Team decides what is needed to deliver the Sprint Goal. They plan from the Product Backlog, starting with the highest priority items, and through discussion with the Product Owner, until they have created the Sprint backlog. The Sprint Backlog can be changed by the Development Team as it "learns more about the work needed to achieve the Sprint Goal." - either by adding or removing items. The Scrum Guide goes even further in stating: "Only the Development Team can change its Sprint Backlog during a Sprint."

If the Product Owner wants to change the direction or goal of the Sprint once it has started, then the Sprint should be cancelled and re-planned. It is fine to add PBIs to the Product Backlog at any time though, including any extra functionality or suggestions that arise from the Development Team's learnings.

Why is it important to make each story deliver value?

Every product we build is for a customer or end user. If what we build doesn't bring them value, why would they use it?

So, how do we define value?

Value is context specific, and it will mean different things to different people. The key to understanding value is in identifying who the customer is in each case. For an API it may be a client application. For a client application it may be the end user. For infrastructure, it may be a development team, and so on.

The Agile Manifesto (Beck et al., 2001) states "Simplicity–the art of maximizing the amount of work not done–is essential." We interpret this to mean, that by focusing on building only those things that deliver real value to your customer, not only do we ensure that everything we do drives use of our product or service, but also that we reduce time wasted building things that won't be used.

Extending that idea, the 80/20 rule, which grew from the Pareto principle, suggests that 80 per cent of the use of any system comes from 20 per cent of the functionality. Finding that 20 per cent, and therefore where the real customer value is, is the difficult task that faces all Product Owners.

Having said this, Product Owners can only do so much to ensure that the team is always building the most valuable things for their customer. The team, with the assistance of the Scrum Master, will need to focus on slicing work so that they can deliver that value early in development, and frequently thereafter.

Can you carry PBIs over to the next Sprint?

Ideally, there would not be any items to carry over at the end of the Sprint, as the aim should be to complete all the PBIs that are in the Sprint Backlog. Work items taken into the Sprint should be broken down sufficiently so that the team is able to complete it in a single Sprint.

However, we work in a complex environment and things don't always go how we would like them to: sometimes the team is unable to complete a PBI by the end of the Sprint.

If the item is still the most important and most valuable item, then you may want to continue to work on it in the next Sprint. However, all unfinished items should be re-evaluated by the Product Owner and Team during the next Sprint Planning. The item doesn't have to be carried over to the next Sprint if it is no longer important, but it will mean that time working on it would have been wasted.

Scrum gives businesses flexibility. Does that mean a PO can change priorities or direction whenever s/he wants?

No, that is not what it means.

The flexibility that Scrum provides is found between Sprints: at the end of each Sprint, the Product Owner and team can re-evaluate their position and decide where to go next. As opposed to projects where requirements are agreed in detail up front and delivered in one or two large releases, Scrum projects offer flexibility due to their regular delivery and continual evaluation.

In most situations, it makes sense to let the team complete the Sprint, and then reassess your needs and priorities before commencing the next Sprint. Sprints are short so you won't have to wait long.

However, in emergencies or when the work in the Sprint Backlog is no longer needed, it might make sense to cancel the Sprint and restart the planning process. It should be very rare that this happens and should be avoided as it interrupts the team's flow and will reduce productivity. If you find it occurring regularly, then you should investigate to find the root cause (for example, maybe your market is highly volatile, your Product Owner needs help establishing requirements, etc).

What is an epic?

'Epic' is not an official Scrum term, but it has been adopted by many teams to describe a type of Product Backlog Item: one that is too big to complete in a single Sprint.

Example epic: "As a customer, I want to be able to have wish lists so that I can come back to buy products later."

It is quite common for low priority items in the Product Backlog to be left as epics, because it rarely benefits a team to discuss in detail work that won't be started for a long time - or may never get actioned. When an epic gets close to being worked on, it will need to be broken down into smaller PBIs that are independent of one another so that any one can be taken into and completed within a single Sprint. This approach helps support core tenets of the Agile Manifesto of delivering small increments frequently.

Some teams break PBIs down into tasks so they can be assigned to individual team members. This is not an approach we like because it encourages teams to split work horizontally rather than vertically, and doesn't promote teamwork. There are also other terms (such as themes, sub-epics) that are used by various software tools or scaling-frameworks. These often confuse teams so we don't recommend using them.

How do we split up a story that is too big?

If you are finding that some items in your Sprint Backlog aren't being completed within the Sprint, then it could be because your PBIs are too large. Having small, achievable chunks of work help you to be efficient and complete items regularly.

You need to find the fine line between PBIs being too big and being too small: they should be big enough to deliver something valuable in their own right.

There are many ways to split work up. We suggest that you work on the simple, core features first because these will deliver most of the value and learning. Then, depending on your type of business, you can split work up into other smaller chunks.

Can you produce a simpler version to begin with? What is the minimum you need to build for the feature?

Are there different functions being performed? For example, can you split the PBI into separate parts for entering information, editing it, exporting it, etc?

Are you serving multiple types of customers? For example, can you split the PBI according to geographical location, spend levels, payment methods, professions (you may have differing functionality for the general public as opposed to suppliers or those within the trade)?

Do some areas need more investigation and experimentation? You may want to start by working on the parts that will give you the most knowledge, so that you are able to approach and complete the rest quicker.

However, you choose to break down your work, remember to make sure that every PBI delivers something of value to the customer even after it has been split.

Should you split stories down into tasks?

Splitting a PBI into tasks is a bit like breaking a meal into ingredients: the individual components give little value in themselves, but together they make something wonderful for the customer. However, some teams find working in this way helps.

If you are splitting items in to tasks, the popular approach is to make them 'SMART' - although what the acronym stands for seems to be hotly debated!

Bill Wake (Wake, 2003) says that SMART tasks are:

- Specific: detailed enough so that everyone understands it
- Measurable: you need to decide how you are able to judge when the task has been finished
- Achievable: someone has to be able to complete the task in a Sprint
- Relevant: it should contribute to a PBI which has value to a customer
- Time-Boxed: not necessarily in a unit of time (for example, hours or days), but you should have an idea of roughly how long it should take to complete the task. Wake says, "there should be an expectation so people know when they should seek help."

Personally, we don't tend to encourage splitting PBIs into tasks because it encourages horizontal slicing, rather than delivering value to the customer as you go along.

What are vertical and horizontal slicing?

Sometimes people request features that are large, complicated, and involve building many components (such as a front-end user interface, a database, a back-end administration function, a mailing system, and so on). You will usually have to split such requests up, delivering small chunks incrementally. But how?

It is tempting to take one of the individual components (for example, the backend database) and work on it until it is finished, then work on another individual component (such as the backend administration function) until that is complete, then take another individual component... until all the individual parts are done. This is 'horizontal slicing'. You work on each component separately and in a linear approach.

Alternatively, you could complete a bit of each component necessary to give a user a basic version of the feature (for example, you might release a web page that allows a user to enter a name, and build a database that holds that single item of data, and produce an administration system that allows you to view the entry). This is 'vertical slicing'. You work on all components at the same time, releasing small increments each Sprint.

We prefer the vertical slicing approach because the user gets something useable early on rather than having to wait until everything is completed. In our example, being able to enter one field on a web page gives the user little functionality. However, it is better than a fully completed, functional database because that doesn't give them any value at all.

This approach also reveals problems earlier because you don't have to wait until everything is finished and released in a big-bang style deployment: you will see issues as you build a little bit more functionality each Sprint.

A great exercise to help people see the benefits of vertical slicing is Alistair Cockburn's Elephant Carpaccio (Alistair.cockburn.us, 2008)

How do we prioritise work?

The Product Owner, who represents the customer, does initial prioritisation.

The Product Backlog should be prioritised by the Product Owner, based on the benefit each PBI would provide to the end user/customer once completed and implemented (and the cost that would be incurred by not doing it). This is why it is essential that the Product Owner has a good understanding of the business and has a position of authority to make such decisions.

The Product Owner may change the priority of a PBI once it has been discussed with the Development Team. However, we discourage Product Owners from relegating items down the queue just because they are big; instead, we encourage teams to break these big, important items into smaller chunks. This way, the most valuable features are still implemented. We would also discourage deferring higher risk items, because tackling these items early on helps remove uncertainty and helps you with future direction.

What is the difference between a burn up and a burn down chart?

The Burn up chart shows daily progress towards a goal by plotting how much work has been done (y-axis) against time (x-axis). As the chart also displays the total scope of the Sprint Backlog as the goal, it effectively highlights any change in scope (because the line representing scope will rise up) which will have an effect on a team's ability to deliver on the goal.

A Burn down chart shows how much work is remaining in the Sprint (y-axis) against time (x-axis). Perhaps the Burn down chart is the better choice to communicate Sprint progress, as it will give you a good view of how much work you have left to complete within the remaining time.

In reality, they are pretty much interchangeable and it's down to personal choice to use one over the other, or either of them at all.

Here are some basic examples. There are many ways to adapt these to display more data points, so have some fun experimenting.

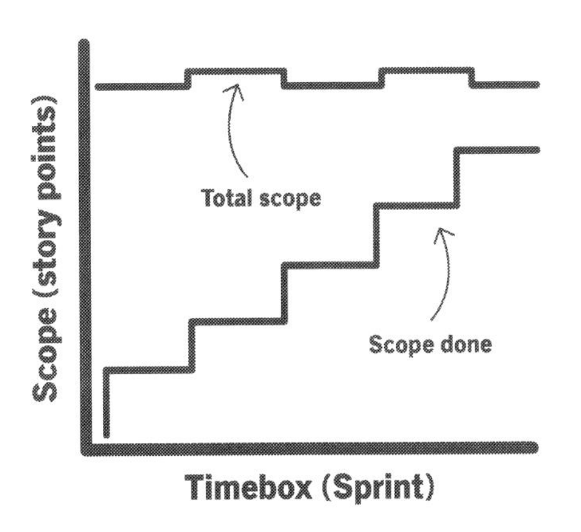

Above: example of a Burn up chart

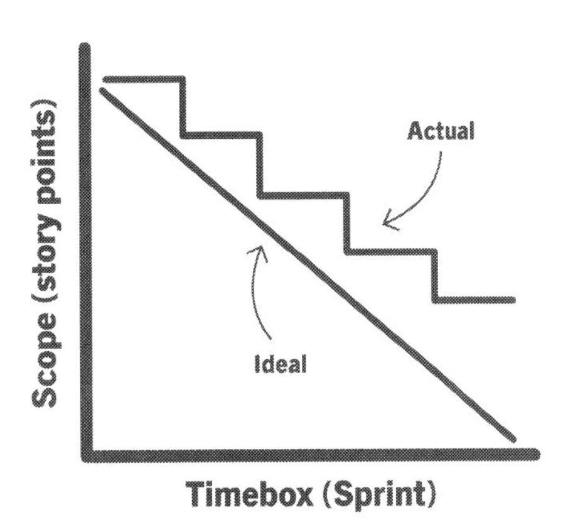

Above: example of a Burn down chart

What is a Scrum board?

A Scrum board (aka 'Agile board') is a way of visualising a team's current position throughout the Sprint. It is also useful as a planning tool within the Sprint.

Boards comprise multiple columns, with the columns reflecting the various stages an item of work goes through before it is completed. There is no limit to the number of columns a board can have (as there should be as many columns as necessary in order to best visualise the team's system), but there will be at least three columns showing:

- Items that have not yet been undertaken (often entitled 'Backlog' or 'Not Started')
- Items that are currently being worked on (often entitled 'In Progress' or 'In Development')
- Items that have been completed (usually entitled 'Done')

The middle column (items that are currently being worked on) is the one that is frequently split out into further columns.

A card is placed on the board for each item of work that is in the Sprint. Some teams display stories, others break stories down into tasks. As each item of work progresses through the system, the card on the board should be moved to reflect its position. Some teams move the cards at the daily stand-up, others move them in real time. Our preference is for real time, but it is up to each team to decide how they update their board. The important point is that the board reflects reality, not what you would like reality to be.

Visualising your work allows the whole team to plan how they will tackle the work in each Sprint, as well as see any bottlenecks that might be forming.

Boards can be physical (using sticky notes or index cards stuck to a whiteboard, wall, window, etc), or electronic.

Do I have to use a Scrum board?

As with most of the practices and techniques described in this book, context is key. We believe that you should only use tools, practices and processes that suit your circumstances and add benefit. Having said that, we've not found a single instance where visualising your work doesn't benefit the team and your business partners.

The Scrum Guide™ (Schwaber and Sutherland, 2016a) states that transparency is important: "Significant aspects of the process must be visible to those responsible for the outcome. Transparency requires those aspects be defined by a common standard so observers share a common understanding of what is being seen."

Visualising your work provides clarity, removing the stress of not knowing what's going on. Every software Development Team we've ever worked with has more work than they can possibly do at any one time. The board serves a dual purpose for those teams. Firstly, it's a planning tool. Teams can see what is in the Sprint Backlog and use the board to define how they will approach the work. For teams with a more fluid process, work can be prioritised on the board by the Product Owner so that the team can easily see what's coming up. Secondly, the board also acts as a communication tool: stakeholders can come to the team area and quickly find out exactly what's being worked on, and by whom.

Physical vs electronic boards

Of course, this is down to personal taste and circumstance.

Physical boards are very popular for a variety of reasons:

- They are very cheap to set up: you just need a bit of wall space, some cards or sticky notes, and maybe a bit of tape to define the columns
- It is very quick to add new requirements, reprioritise them and remove redundant ones
- There is something satisfying about the act of moving a paper-based PBI across the board. It is also a very public display of progress.

But electronic boards have their benefits too:

- Team members who are working from another location (whether occasionally or permanently) can still see and interact with the board
- Some of the software available is very clever and can save you time by running real-time reports (although we recommend you ask yourself what you are going to do with these reports before you get too carried away)
- You can usually add notes and attachments to PBIs, thus centralising information and making it easy to see the history of an item as well as its current position.

Should my board have swim lanes?

Swim lanes refer to horizontal rows on a Scrum board. Many boards do not have swim lanes, but some teams use them to identify different projects, different types of work (for example, bugs and features), but most commonly they are used when a team breaks PBIs down into tasks. In the latter instance, each PBI will have a swim lane, with a card displaying the PBI in a far left column, with all other columns holding tasks relating to that PBI.

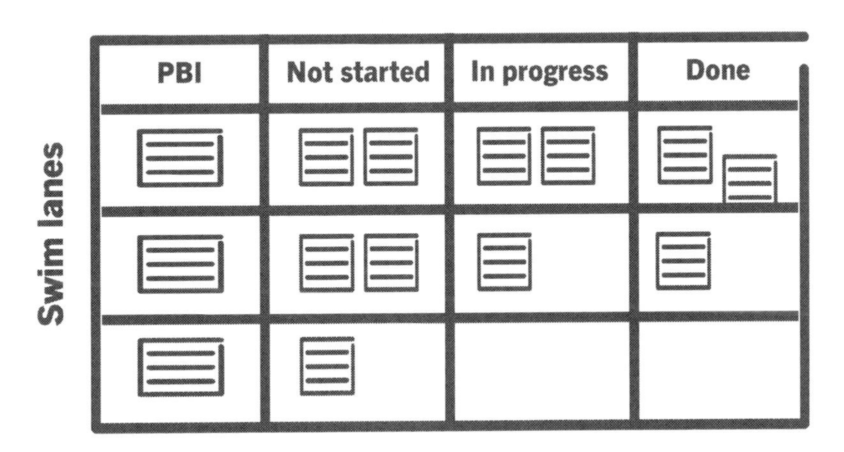

Most teams will have separate boards for different projects. However, using swim lanes to identify different projects can work, particularly if you have people shared across both projects (i.e. someone has to work in both swim lanes). In general, we advise that teams focus on their specific work and that you don't split your people across different work streams. The important question to ask when considering using swim lanes is: how will we know which item on the board is the highest priority? If you can answer that, then maybe swim lanes are right for you.

For example, maybe you want a swim lane for urgent live bugs. You might agree that you always pick up a live bug before starting a new PBI, or maybe you have a rota of people responsible for any live bugs that arise. You may want to highlight any work that has stopped you from working on your Sprint Backlog, but that doesn't need to be done using a swim lane.

What are Conditions of Satisfaction?

The description of the PBI will tell you who it is for, what it is and why you are doing it, but will not tell you everything; it is the starting point for a conversation.

Before you start work on the PBI, the Product Owner will often be able to tell you some of the criteria for considering the item a success; others will emerge as the team and Product Owner discuss the requirements. Collaboratively, the team and Product Owner will produce a list of criteria for the PBI, which is what we refer to as the Conditions of Satisfaction.

They are specific to a PBI and should not be confused with the 'Definition of Done'.

The Conditions of Satisfaction are not a definitive list of requirements; they are just some of the known criteria that are desired. The team and Product Owner should discuss the PBI when needed rather than rely on the Conditions of Satisfaction.

Are Conditions of Satisfaction the same as Acceptance Criteria?

Most people seem to use these terms interchangeably, but we think there is a subtle difference.

Conditions of Satisfaction (often written as CoS) will usually originate from the Product Owner when a PBI is presented to the team, and help describe what the user wants.

Acceptance Criteria, although based on the PBI and Conditions of Satisfaction, are usually generated by the team to describe factors that would cause a test to pass.

A very simple example:

PBI description
As the marketing manager of ABC, I want to know how many visitors we have to our different websites, so that budget can be allocated according to demand.

Conditions of Satisfaction
- Use existing Google Analytics accounts for company Y.
- We do not need to record users visiting the company Z site.

Acceptance Criteria
- Given that I'm visiting the company Y website
- And that I've not visited any page on the website within the last 30 days
- When I reach any page of the website
- Then Google Analytics will record a unique user against account UA12345678-9

Because Acceptance Criteria can be used for various purposes, if you are talking about factors that the business wants fulfilled as part of the PBI, then stick to using Conditions of Satisfaction. It seems that its meaning is used much more consistently.

Please note that Given, When, Then is not the only approach to writing Acceptance Criteria.

What are entry criteria?

Entry criteria are similar to exit criteria. However, while exit criteria define what must happen to a PBI so that it can be pushed into the next column on your board, entry criteria is defining what must happen to the PBI 'before' it can be 'pulled' into the column.

As someone will only pull a PBI into each column as they are able to work on it, this prevents a backlog forming. When work is being pushed, there is nothing stopping people just pushing more and more work downstream.

Teams using the Kanban Method prefer entry criteria, but there is nothing to stop you using them with Scrum.

An example of entry criteria for an 'In Development' column could be:

- Conditions of Satisfaction defined
- Agreement made on testing requirements
- Shared understanding of the PBI
- PBI has been sized

What are exit criteria?

Exit Criteria are similar to a Definition of Done, but for each column on your Scrum board. They tell the team what must happen before a ticket can leave one column and be 'pushed' to the next. They are like checklists for each stage of your process.

They can list as much detail as you wish. For example, before a ticket in the 'Passed Code Review' column could be pushed to the 'Merged' column, one of the teams I worked with stated that the following must be completed:

- Merged to correct environment branch, tagged and RPMs built
- Change log for RPM build verified

The responsibility for checking that the criteria have been met is with the person moving the PBI (or task) to the next stage.

Estimation

How Long will it take? How much will it cost? These chapters discuss the basics of planning poker, velocity and release planning and give clarity to some of the more nuanced parts of the role estimation plays within the Scrum framework.

Why do we now 'forecast' instead of 'commit' to items in a Sprint Backlog?

There has been much discussion in the Agile and Scrum communities about the use of the terms 'commit' and 'forecast' in relation to the Sprint Backlog. Originally, teams were said to 'commit' to complete the items in the Sprint Backlog; now the team is said to forecast completion of the same items. The use of the word commit was intended to motivate and get buy-in from the whole team.

However, in reality, many stakeholders took the term more literally and assigned a strong sense of failure to teams that did not complete all the items in a Sprint Backlog during a Sprint: estimates had become deadlines and promises. As a team, we can only do our best to try to complete what we estimate is possible during the time available. Software is a complex domain that has many unknown elements.

To solve the misunderstanding, the community dropped the use of the word 'commit' and embraced 'forecasts' (including an official change in the 2011 edition of The Scrum Guide™ (Schwaber and Sutherland, 2011)).

If you want to use 'commit' instead of 'forecast', please do so; just make sure that everyone understands what you mean.

Why do we estimate in points?

Relative sizing is easier than providing an exact figure.

Mike Cohn, in his book Agile Estimating and Planning (Cohn, 2006), compares it to ordering food and drink. When ordering a drink, you normally don't ask ask how many millilitres of fluid the medium-sized drink contains. You normally don't ask how many fries are in a medium-sized portion or how many grams it weighs. You use past experiences as a guide and order accordingly to satiate your thirst and hunger.

We do the same when estimating work: we compare a new requirement against other items, based on a mixture of complexity, effort and risk (and any other important factors that are appropriate to your type of work).

However, what if you're just starting the project and you have nothing to compare against? In this instance, you will need to select a PBI to use as a baseline. Some teams pick one of the smallest requirements and give it the value of 1, other teams choose a medium-sized item and call that a three- or five-point PBI. All subsequent items are then compared to that initial, and all subsequently sized, PBI. The initial assignment of size is arbitrary: just don't pick a medium or large story and call it a one-point PBI because you can't give smaller items less than one point (we don't want to start sizing in fractions!)

Cohn uses the example of dogs to explain the process of relative sizing. If I give you a list of five dogs (for example, terrier, poodle, Labrador, spaniel, Great Dane), you should be able to give me an idea of their relative sizes. Although there are a number of questions that you will no doubt want to ask about the dogs (for example, "what type of poodle is it?"), you could still estimate their relative size without further information. It would be a lot harder if I asked you to tell me their height in metres.

This explains why people involved in 'knowledge work' prefer relative sizing: even though the team rarely has all the information they would like, they are still able to guess a requirement's relative size compared to previous work.

Why do some people estimate in days?

Although many teams estimate in points, both Mike Cohn (Cohn, 2006) and James Grenning (Grenning, 2002) refer to sizing in days. This doesn't mean actual days (or 'elapsed time' as Cohn refers to it), it means days that it would take if you lived in an ideal world where you had everything you needed to complete the work, where your infrastructure worked as required, where people were available at your demand. This is why it is commonly referred to as an estimation of 'ideal days'. This time-based figure can then be used to calculate velocity of the team similar to how story-points are used.

Like story-points, ideal days estimations are not looking at each individual's input on the item; it's the sum of all the time it will take the team to completely finish it. For example, how long does it take to bake a cake or put up some shelves? The actual time cooking the cake only takes 20 minutes and the preparation time is at least double that, but even that estimate doesn't include the time spent shopping for the ingredients or allowing for the cake to cool down. Similarly, putting up a shelf might only take 30 minutes, but it can take hours if you include time spent selecting and purchasing the fixings, filling old holes and cracks then sanding and repainting walls, drilling new holes, dealing with problematic plasterboard and tidying up. It is asking, "How long will it take to complete the item even if everything goes well?"

Is estimating using story points better than estimating in ideal days?

These methods are very similar because neither is actually using 'elapsed' time.

Ideal days estimates how long something would take if the world was exactly as you wanted it to be. For example, you had everything you needed to get the job done, everything worked as expected, and there were no interruptions.

Sizing using story-points uses arbitrary numbers to estimate the size of something relative to other items that have been completed previously.

Both use 'velocity' to enable a team to estimate how much future work will be completed in a Sprint or to estimate how many Sprints it will take to complete a project. Velocity is the average story-points (or average ideal days) of the PBIs that the team completed in the previous Sprints.

Story-points is currently more popular, probably because it removes the potential for misunderstanding between ideal days and elapsed days.

What is Planning Poker® / Scrum Poker?

The aim is, as James Grenning (Grenning, 2002) described it, "to avoid analysis paralysis". The concept, which comes from Extreme Programming, is used during release planning with the aim of getting a "ballpark estimate of the effort to build the product... precision of individual estimates is not the goal. Determining the scope is." It acknowledges that these are estimates and that some will be under-estimates, whilst others will be over-estimates.

This process is meant to be quick and involves the whole team:

- Someone (often the Product Owner) explains the requirements
- The Development Team asks any questions necessary to gain an understanding of what is desired
- Each member of the team individually comes up with an estimate
- Everyone declares their estimate at the same time (hence the word "poker" in the name)
- If everyone is in agreement, then the estimate is recorded and the team moves on to the next PBI
- If there is disagreement, which is normal, then the people who declared the lowest and highest estimates explain their reasoning. After a bit of discussion, everyone provides another estimate. This process continues until consensus is reached.

The most important part of the whole process is the conversation, not the number you come up with. Everyone's input is important - the newbie developer on their first day might have a good point that you've all missed - and the discussion around requirements often reveals a lot. It is especially useful in pointing out when there is a significant disagreement for the effort involved and often highlights when requirements have been misunderstood by some members of the team.

The term Planning Poker ® has been registered as a trade mark by Mountain Goat and promotes the use of Fibonacci-like numbers: 0,1, 2, 3, 5, 8, 13, 20, 40 and 100.

Why the Fibonacci series? Why not 1, 2, 3, 4, 5, 6, 7, 8, 9..?

Whether you are sizing/estimating using unit-less numbers or ideal days, a higher number represents a larger item of work.

Both James Grenning (Grenning, 2002) and Mike Cohn (Cohn, 2006 recommend using number sequences that have gaps as the numbers get larger (1,2,3,5,7,10,infinity and 0,1,2,3,5,8,13,20,40,100,infinity respectively). But why?

As the estimates get larger (or longer), there is less benefit in being precise. If you were able to choose any number between 1 and 100, could you explain why you had chosen the PBI to be 49 points instead of 48? Remember, you are estimating, not providing precise timings or deadlines.

This is why Grenning uses infinity once you go above 10 points (or 10 days) and Cohn simplifies the Fibonacci series (which generates numbers by adding the previous two numbers together, eg 3 + 5 = 8) to use 20, 40 and 100 rather than 21, 34, 55 and 89. The effect is the same: if your number is above 10, then you need to be breaking the item of work down into smaller pieces.

Is a 5-point PBI twice as big as a 3-point item?

No, a 5-point PBI is smaller than two 3-point PBIs.

Mike Cohn (Cohn, 2006) uses a good metaphor to help explain:

Imagine you have a number of buckets of various sizes: a 1 litre bucket, a 2 litre bucket, a 3 litre bucket, a 5 litre bucket, a 8 litre bucket, a 13 litre bucket. If I give you 6 litres of water, what is the smallest bucket you could put it in without losing any? It has to go into the 8-litre bucket as the 5-litre bucket would be too small.

The same thinking applies to PBIs. If you think something is twice as big as a three-point item of work, then you would give it a size of eight.

Why do you need consensus in sizing sessions?

Consensus is encouraged because everyone's view is worth hearing. Everyone (including junior and new members of the team) have different perspectives, fresh ideas and may think of simpler solutions.

Don't dismiss the contribution everyone on your team can make, regardless of their experience or time on the team. Without the need for consensus, it is easier to dismiss people's opinions and allows strong characters to dominate discussions. Pushing for consensus gives people the chance to explain the reasoning behind their estimate and an opportunity to highlight something that others might have missed.

However, don't be too strict on enforcing consensus: you are only trying to reach an estimate. For example, if you've discussed an item of work and all agree what the requirement entails and that it is relatively small, then there isn't much sense in spending a long time discussing whether it is a one- or two-point story. Grenning (Grenning, 2002) suggests that you discuss disagreements over estimates, but acknowledges that, "If you can't get consensus, don't sweat it... Defer the story, split it, or take the low estimate."

Some teams prefer to take the median or even the higher number. Implement whatever approach makes you happy - just make sure you keep your approach consistent.

Does sizing help define requirements?

The process of sizing and estimation (including the Planning Poker ® method) is not just about coming up with a number to reflect the size of the PBI; an important part of sizing is the discussion that precedes any estimation being given.

The Product Owner brings the requirements to the meeting, whilst the team comes up with the solutions. To reach this point, the team asks the Product Owner questions about the PBI and everyone discusses the requirements until the team understands what is needed. Sometimes, the solution offered might cause the Product Owner to reconsider the request that was originally presented. For example, a request that the Product Owner expected to be a 'quick win' might be a complex request or not be the best solution for what the users actually want. Some requests can result in important discussions about what is really required before any estimation is provided by the team.

So, yes, Planning Poker ® and other methods of sizing can help define requirements because of the discussion that occurs.

What is a spike? When should they be used?

The term 'Spike' was first coined by Kent Beck. In Extreme Programming Explained (Beck, 1999), Beck referred to a spike as a quick "architectural sketch" that would "drive a spike through the entire design", which would then provide clarity to the team on a particular problem.

In essence, it is an investigation into an area where further knowledge is required in order to progress the piece of work. This could, for example, include how to break a large requirement down or an investigation into how to implement a requirement if the solution is completely unknown. The spike doesn't need to result in shippable functionality, and all focus should be on gaining the knowledge to progress the PBI.

Spikes should be time-boxed and should be relatively short, ideally no longer than a day or two. It is up to the team as to how long they feel they need for the spike.

Spikes should only be used when you have absolutely no idea how to approach something and should not be used to result in a perfect specification document. Teams should aim to work on and release functionality wherever possible.

Why don't we size spikes?

The goal of a spike is to find out information and gain knowledge, not to deliver 'working software'. A spike does not deliver anything functional to the user; if any output is generated during this process, it is often disposable.

So, if you gave these spikes story-point values and counted them within your velocity calculations, your velocity would not be a true reflection of the amount of 'working software' delivered in a Sprint.

Instead, spikes are time-boxed: you stop working on them when the time allocated to the experiment has run out.

We recommend:

- Keep spikes small so that the time-box is short; time spent on spikes means you aren't spending time producing working software
- Use spikes sparingly and don't run multiple spikes in a Sprint
- Don't over-commit; take spikes into account when you are calculating how much work to take into a Sprint Backlog

What is velocity?

Velocity represents a team's average rate of progress. It is the average number of story-points that the team completed over previous Sprints.

$$\text{Velocity} = \frac{\text{Sum of completed story points}}{\text{Number of Sprints}}$$

Teams can use velocity as a guide for how much work they might complete in the next Sprint.

For example, if a team completed the following points per Sprint then they would have completed a total of 64 points over four Sprints. The team's velocity would therefore be 16 (64 / 4 = 16). The team would then use 16 points as a guide to forecast work for Sprint 5.

Sprint	Points completed
1	10
2	15
3	20
4	19

Mike Cohn, in Agile Estimating and Planning Cohn, 2006), says, "A key tenet of agile estimating and planning is that we estimate size but derive duration."

This means that we can use velocity to estimate the duration of a project. To do this, take the sum of story-points from all items in the Product Backlog, then divide by velocity. The result is the number of Sprints that it will take to complete the project (that is, all items in the Product Backlog).

A benefit of using velocity is that it takes account of improvements in a team's performance, so you don't need to re-size PBIs in the Product Backlog as the team improves (because the team's velocity will increase as the team delivers more points per Sprint).

Do not use Velocity to compare different teams' productivity because it is intrinsically linked to story-point values given by each team (for example, what team A considers a three-point story could be considered an eight-point story by team B. This would result in team B having a much higher velocity even if they completed exactly the same work).

Is velocity a guide for the business to judge the team's effectiveness?

No.

Velocity is meant as a guide to help the team decide how much work to take into a Sprint. For example, if a team is completing an average of 30 points per Sprint, then they should take on about 30 points in the next Sprint; they should not accept 60 points worth of PBIs!

Velocity is not meant to be used to compare teams. Story-points are relevant only to the team that created them. Two teams discussing the same PBI might give it totally different points. One team might call it a two-point story whilst the other gives it five points. It isn't delivering more value to the second team and isn't a bigger piece of work either, the teams are just using a different scale. Even if both teams completed the same PBIs, the first team would have a much lower velocity because they are using smaller numbers.

Velocity is not meant to judge a team's performance either. If a team completes far fewer points in a Sprint than their velocity, it doesn't mean that they were being lazy; ask them what problems they had and try to help them to stop it happening again (for example, are there problems with the environment, their processes, unclear requirements?) Remember, velocity is an average of what the team has been delivering (although it's not good if a team delivers no points in a Sprint, then 60 points in the following Sprint).

Don't judge a team based on points completed or their velocity; ask how you can help to make things better.

Why don't you count some story-points when partially completing a PBI in a sprint?

Imagine that you get to the end of the Sprint and one of the eight-point PBIs is nearly finished. Why can't you count five points towards the Sprint's velocity, and carry over the remaining three points to the following Sprint?

Firstly, it is difficult to know how many points to take because you never really know how close you are to actually completing something. We've heard people say that a PBI is "90 per cent complete", but the last 10 per cent takes longer than it took to complete the first 90 per cent.

'Done' is binary: it's either done or not. Velocity is calculated using an average of recent Sprints, so it doesn't really matter which Sprint a PBI's points are attributed to.

However, we do not advocate holding back something that has value and is production ready. If you have completed part of a PBI, and that part is functional in itself and can be released, get it live and allow your users to start benefiting as soon as possible. In this situation, we don't see the harm in splitting the original PBI into smaller chunks at the end of the Sprint, releasing the part that is ready, then re-evaluating the remaining smaller PBIs.

If you find that PBIs are frequently nearly finished at the end of the Sprint, consider splitting PBIs into two or more smaller PBIs before the Sprint starts (with points divided accordingly).

When should we re-size?

To define when we could re-size work items, it may be useful to identify when there is no need to re-size.

When work is started but not completed during a Sprint that work will roll over to the next Sprint. There is an argument to be made that you could re-size the remaining work at this point, but given that velocity averages out over time, there really is no need. The work done on the uncompleted work items does not count towards the story-points completed, and all story-points move forward to the next Sprint.

There are two scenarios where you may want to consider re-sizing work items:

- If work has been started and new information comes to light that dramaticlly changes the solution, then the size will have been affected. If we've learned our original assumptions don't hold, continuing to work to the original size makes no sense, and re-sizing the work seems appropriate.
- If significant changes are made to the make up of the team, then re-sizing work may also be appropriate. Teams learn how to size together, and if several teams members leave, and, or a number join, the new team will likely size things differently.

How do you deal with estimates when the team has different skill levels?

Estimation using story-points considers the overall size of a PBI (based on how big or complex it is) relative to others, not how quickly someone could complete it.

For example, if you already have a PBI that you have sized as onepoint, and you consider a new requirement to be twice as big, then the new PBI will be sized at two points. That the new two-point PBI might take Alex five days to complete, but would only take Bob three days, is irrelevant; it is still a 2-point PBI. Velocity deals with varying skill levels by looking at the average output of the team over time.

Should the Product Owner and Scrum Master be involved in sizing?

The Product Owner is certainly involved in the estimation process because it is the PO who supplies and clarifies requirements to the team. As scrum.org says in The Scrum Guide™ (Schwaber and Sutherland, 2016a), "The entire Scrum Team collaborates on understanding the work of the Sprint".

However, it is normally just the Development Team that supplies estimates during sizing; as POs and Scrum Masters are not going to be doing the work, they do not usually contribute.

Armed with the Development Team's estimate, the PO may choose to amend the PBI by changing the scope, de-prioritising it or removing it from the Backlog completely.

What is the Iron Triangle?

Dr Martin Barnes, founding member and former president of the Association for Project Management (APM), proposed the original concept of the iron triangle back in 1969 (Weaver, 2007).

The iron triangle is a metaphor to explain the relationship between four factors of product development: quality, time, scope and cost.

Quality is in the centre of the triangle because it is a non-negotiable factor.

The relationship between the three other factors, positioned at the corners of the triangle, dictates that a maximum of two can be fixed: at least one must be negotiable.

For example:

- You can get everything you want (scope) and quickly (time), but you may need to pay more for it (cost); or
- You can get everything you want (scope) and cheaply (cost), but you need to be prepared for it taking longer than you'd like (time); or
- You can get something quickly (time) and cheaply (cost), but may not get everything that you want (scope) within that timescale and for that cost.

How do we set deadlines using Scrum?

If time is what is most important, then story point estimates and velocity should give everyone a good idea whether a team can deliver a certain amount of scope in a given time. Please see the relevant chapters for an explanation.

Underneath everything, setting deadlines for a team working within the Scrum framework is no different from the way you might do it for any other team. Although defining a time box is the traditional method for setting deadlines, they can come in a number of forms of which time is just one.

There are three main factors to consider within any software project: time, cost and scope. The general principle is that you cannot fix all three of those things, and at least one has to be flexible if you are to maintain the quality of the resulting product. 'Fixed everything' projects are not uncommon, but eventually something will flex. The question is what is most important to the client or the stakeholders of the project: the time it takes to deliver, the scope or features delivered, or the cost of delivery?

More than likely you have either quoted your client or you have a Development team that's a fixed size and that can't easily grow. The probable deadline setting scenarios are therefore as follows:

- Cost and time are fixed, scope is flexible: Use the product backlog to prioritise and set a cut-off point. Deliver the higher value features first, and if the high value features create more valuable work, prioritise that against the remaining backlog.
- Scope is fixed, time and cost are flexible: The old adage that time is money holds true. In all cases, taking more time will cost more money, whether the team size is fixed or not. In this scenario, continue to prioritise as in the first scenario, and simply keep working until you're done.
- Scope and time are fixed, cost is flexible: This probably sounds too good to be true, and in fact it is. If you do add people, try doing it at the start of the project. Adding people part way through will cost you time as they are onboarded.

Above all else, the team must work closely with the Product Owner and stakeholders to find out what is more important. The Product Owner must revisit the backlog frequently, and the team must relay what they have learned regularly to ensure that collectively, they continue to head in the right direction.

What is a Burn Down chart? How do I build a Burn Down chart?

A Burn down chart aims to give the team visibility on the amount of work remaining in a Sprint. This can be shown using either:

- Total time needed by the team to complete the unfinished PBIs in the Sprint Backlog; or
- Sum of story points of the unfinished PBIs in the Sprint Backlog

To create a Burn down chart using time:

On the y-axis, plot the sum of the estimated time that it will take the team to complete all the PBIs in the Sprint Backlog. You can use days or hours. On the x-axis, plot the length of the Sprint in days. Draw a straight line from the sum of time on the y-axis to the end of the Sprint on the x-axis. This line is the guide of how much work should be outstanding as you go through the Sprint.

Then, each day, plot the sum of how much time is still outstanding (i.e. how much time the team needs to complete all the items in the Product Backlog). Please note that you are not calculating the time spent working on items because this might not correlate with how much time is remaining. As you go through the Sprint, you will find that your initial time estimates were over or under, so it is important that you record the amount of time left.

If you find that you are above the guideline, then this suggests you are not going to complete everything in the Sprint Backlog before the end of the Sprint. If you are below the guideline, then you may finish the work before the end of the Sprint.

To create a Burn down chart using story points:

Follow the above instructions, but replace time with story points (i.e. what is the sum of story points for all PBIs in the Sprint Backlog that have not been completed?)

Should we use time or story points to build a Burn Down chart?

Some people say you should use points, others argue for time.

An argument against time-based Burn down charts is that they often appear to show that the team is on target to complete all the PBIs in the Sprint Backlog, but then the team makes a discovery that increases the time needed by many hours. Accurately judging time for knowledge-based work is very difficult.

An argument against point-based Burn down charts is that our goal is not to complete as many points in a given Sprint (this can easily be gamed by just increasing the size of stories in planning), it is about delivering benefits to our customers. We can see what has been completed by looking at which PBIs are in the 'Done' column.

However, we do not know which metrics your team and stakeholders find beneficial; use whatever works for you.

What is a Definition of Done?

A Definition of Done (also known as a DoD) is a concise list of criteria that defines the minimum quality standard that must be met by each PBI within the Sprint. A PBI is only 'Done' (that is, finished and potentially shippable) once all of the criteria in the Definition of Done have been satisfied. It is important to understand that this is focusing on the quality of all work, not just the functionality of the individual PBIs.

The whole team must collaboratively understand and agree the criteria that make up the Definition of Done. This ensures that there is transparency, allowing the team to work towards their goals using the same approach and rules. A Definition of Done varies per team (although, if multiple teams are working on the same product, you may benefit from mutually agreeing a standard DoD for all teams).

You should expect a Definition of Done to change as the team matures, the product develops and a higher quality of work is strived for. However, changes are made between Sprints, not during a Sprint.

An example Definition of Done:

- Every PBI must have been code reviewed by at least two people
- Every PBI has been functionally tested
- All unit tests are passing and the build is green
- All code has been deployed to the production environment

What is a Definition of Ready?

"In a suitable state for an action or situation; fully prepared" is the Oxford Dictionaries' (Oxford Dictionaries | English, 2016b) definition of 'Ready'.

Many successful teams agree on criteria that must be met before a Product Backlog Item can be taken into a Sprint. This increases the chances of the team completing the item in the Sprint and achieving the intended goal.

Exactly what the definition of ready looks like varies by team. It is whatever the team decides is the minimum information to get the PBI started. For instance, a team might state that it must be written in a particular format and be accompanied by any known Conditions of Satisfaction.

We find the INVEST acronym useful for defining a good PBI. From Wikipedia (En. wikipedia.org, 2009):

- **I** Independent: it should not need another PBI to be completed before it can be started.
- **N** Negotiable: The scope of a PBI, up until it is part of a Sprint Backlog, can always be changed.
- **V** Valuable: if the work you are doing doesn't deliver value to an end user, why are you doing it?!
- **E** Estimable: there must be sufficient information available to allow the team to estimate the size of the item.
- **S** Sizeable: PBIs should be a good size, meaning that they can be completed within a Sprint.
- **T** Testable: there must be sufficient information to allow the team to know when they have completed the PBI.

Should you limit your WIP in Scrum?

Limiting Work-in-Progress (WIP) is a key step from the Kanban Method, designed to help teams manage the flow of work. It is not a Scrum practice. However, with Scrum, teams aim to release all 'Done' items at the end of the Sprint. We believe that this implies that a team should work on an item, finish it, and then move on to the next. In this way, the team is naturally limiting WIP.

Scrum also kind of limits WIP by having a Sprint Goal and a Sprint Backlog; you shouldn't do work that doesn't get you nearer to the Sprint goal or that isn't in the Sprint Backlog.

In addition to the implicit nature of limiting WIP in Scrum, teams should also consider that there is an inherent cost to context switching. The more work the team has in progress at any point in time, the longer it will take. Ultimately, if you want to apply an explicit limit to the amount of work that the team is undertaking at any one time, this could be considered good practice – just remember that it isn't a documented part of Scrum.

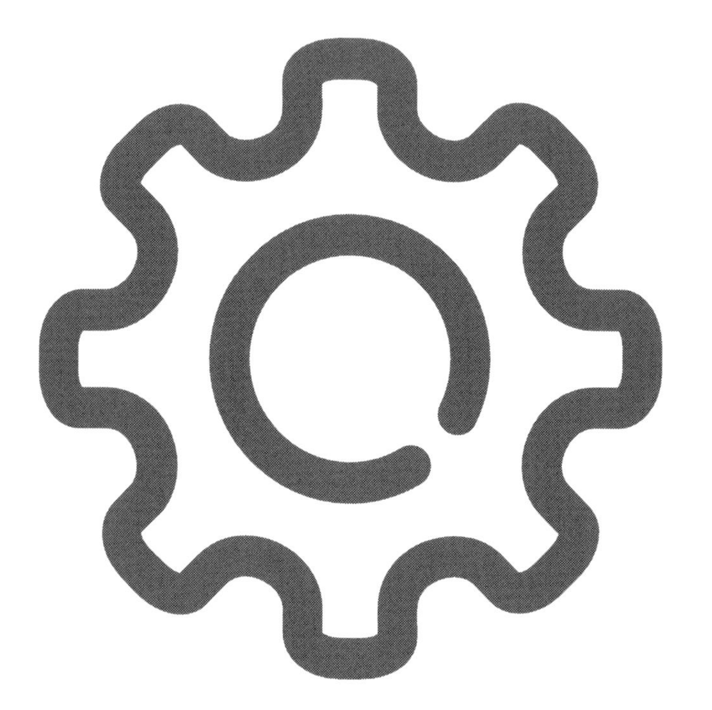

General

These chapters explain some of the other terms that get used frequently but are not directly part of Agile or Scrum. There are also a number of organisations and companies that you will come across frequently, and we've explained their role in the evolution of Agile with Scrum.

Can I break the rules?

Scrum is a framework made up of various components, which support each other, such as roles, events and artefacts. The Scrum Guide™ (Schwaber and Sutherland, 2016a) says that each of these components "serves a specific purpose and is essential to Scrum's success and usage." In other words, Scrum is not modular and removing any of its parts makes the system unstable.

However, how you use the framework is up to you, so it depends on which 'rules' you are thinking of breaking? For example, The Scrum Guide would say that not holding a retrospective is unacceptable, but would be fine with you changing the format of the retrospective. Scrum often highlights dysfunctions in teams and some teams choose to hide the problem by removing part of the Scrum framework rather than solve the real issue. This is commonly known as 'Scrum-but' because teams will be heard saying things like, "We are using Scrum ... but we don't do retrospectives because they are a waste of time."

There is a current trend of changing the names of events, roles and artefacts (usually to avoid resistance from within an organisation). Although this isn't recommended because it can cause confusion, if you have to do it, then at least ensure there is a shared understanding of any variance and that you are able to maintain the principles and values of the Agile Manifesto and Scrum.

What is scrum.org?

Ken Schwaber left the Scrum Alliance in late 2009 and founded Scrum.org with Alex Armstrong. It has a large, global community and expressly focuses on software development.

Scrum.org runs training courses and has various assessments to prove members' skills (both free and paid-for):

- Professional ScrumMasterTM has three levels: I, II and III
- Professional Scrum Product Owner™ has two levels: I and II
- Professional Scrum Developer™ has one level: I
- Scaled Professional Scrum

Working with Jeff Sutherland, Scrum.org maintains The Scrum Guide™ (Schwaber and Sutherland, 2016a) in 30 languages.

In 2015, it created a scaled Scrum approach called Nexus™.

Scrum.org is based in Boston, Massachusetts, USA.

What is The Scrum Guide?

The Scrum Guide™ (Schwaber and Sutherland, 2016a) is a document created by Ken Schwaber and Jeff Sutherland. The first edition was published in 2010, with updates in 2011, 2013 and 2016.

Although hundreds of books have been written on Scrum, we consider The Scrum Guide as the authoritative source of Scrum theory.

The Scrum Guide contains:

- Definition of Scrum
- Scrum Theory
- Scrum Values
- Make-up of the Scrum team (for example, Product Owner, Scrum Master, Development Team)
- Description of Scrum events (for example, Sprint Retrospective)
- Explanation of Scrum Artefacts (for example, Product Backlog)

It is free to access and can be downloaded from Scrum.org.

What is Scrum Alliance?

Scrum Alliance ® was founded by Ken Schwaber, Mike Cohn and Esther Derby in 2001. It is a non-profit, membership organisation with a global network of Agile users, trainers and coaches.

Its aim is to "transform the world of work" - not just software organisations - and states its mission is "to guide and inspire individuals, leaders, and organizations with practices, principles, and values that create workplaces that are joyful, prosperous, and sustainable." (Scrum Alliance, 2016)

Scrum Alliance produces many Agile educational tools and resources, hosts gatherings (from worldwide conferences to local user groups) and has an active online community. However, it is probably most known for its certification programme:

Each of the three entry-level certificates requires attendance of a taught course followed by a short test:

- Certified Scrum Master ® (CSM ®)
- Certified Scrum Product Owner ® (CSPO ®)
- Certified Scrum Developer ® (CSD ®)

Holders of one or more of the entry-level certificates can apply for the advanced level certificate, which is designed to demonstrate experience and knowledge:

- Certified Scrum Professional ® (CSP ®)

Only holders of the CSP certificate may apply for the highest levels of certification:

- Certified Scrum Trainer ® (CST ®)
- Certified Enterprise Coach ® (CEC ®)
- Certified Team Coach ® (CTC ®)

Scrum Alliance has over 450,000 certified practitioners worldwide! It is based in Indianapolis, US.

What is scaling?

In short, scaled Scrum and scaled Agile are when Scrum or Agile processes are applied at 'enterprise' level, with multiple teams working simultaneously to deliver the product.

You may need multiple teams working on the Product if you wish to deliver the items on the Product Backlog more rapidly or if all the expertise required to deliver the whole product cannot be accommodated in one Scrum team.

Scaled Scrum should be underpinned by the values and principles of the Agile Manifesto (Beck et al., 2001) and Scrum, and in most cases has the same events, artefacts and roles. There is only ever one Product Backlog, and generally only one Product Owner. In reality, the Product Backlog may be split in a variety of different ways and some of the responsibility of the Product Owner may be delegated to others, for example proxy Product Owners working with some of the development teams.

There are a number of scaling approaches, and the three most popular approaches currently are SAFe, LeSS and Nexus.

SAFe

Scaled Agile Framework, or SAFe, was created by Dean Leffingwell, with the first version being released in 2011.

SAFe proposes to "provide comprehensive guidance for work at the enterprise Portfolio, Value Stream, Program and Team levels" (Scaledagileframework.com, 2016). SAFe v4.0 claims that it "is appropriate for implementations of under a 100 people to those that require thousands of people" (ibid.)

LeSS

Large Scale Scrum, or LeSS, was created by Bas Vodde and Craig Larman. The framework was formalised in 2013. LeSS is Scrum applied to many teams working together on one product.

They propose that like "one-team Scrum, LeSS is (1) lightweight, (2) simple to understand, and (3) difficult to master — due to essential complexity." (Less.works, 2016).

There are two versions of LeSS. They say that "basic LeSS framework is applicable for up to eight teams, and LeSS Huge is applicable eight teams or more, and can work for up to a few thousand people per product" (ibid.).

Nexus

Nexus was created by Scrum.org in 2015. It proposes to be the "exoskeleton of scaled Scrum" (Scrum.org, 2016). The core tenet of Nexus is to continually identify and remove dependencies between teams who are working on the same product.

Nexus is designed for three to nine teams. These teams must all be working on the same product. If there are more than nine teams then Scrum.org recommends that multiple Nexuses are implemented.

We believe that rather than focusing on scaling Agile processes, you should focus on being Agile. Scaling, even when using these defined frameworks and methods, can lead to many challenges. We recommend trying to keep Scrum teams as independent as possible in order to avoid having to 'scale'.

What does "Minimum Viable Product (MVP)" mean?

The term Minimum Viable Product or MVP, was coined by Frank Robinson in 2001 (Syncdev.com, n.d.), and was later popularised by Eric Ries in his book The Lean Startup: How Constant Innovation Creates Radically Successful Businesses (Ries, 2011).

Ries described MVP as "a learning vehicle". MVP proposes that you learn about your vision by getting potential customers to test an idea, and then adapt your product based on the feedback collected, rather than invest a lot of time and money in a fully-fledged product before you know whether it has the required value. An MVP can take many forms, a survey, a sketch and sometimes even a prototype.

There is a common misconception that an MVP is the minimum amount of work you need to do to a product before you can put it to market. However, in The Lean Startup, he explains "It is not necessarily the smallest product imaginable, though; it is simply the fastest way to get through the Build-Measure-Learn feedback loop with the minimum amount of effort... its goal is to test the fundamental business hypotheses".

What does "Minimum Marketable Product (MMP)" mean?

Many people mean Minimum Marketable Product (MMP) when they use the term Minimum Viable Product.

An MMP refers to the version of a product that contains the fewest features but is considered ready for customers (or at least a subsection of them). In other words, the organisation would not be prepared to release the product to their customers before it has reached this stage of development.

Minimum Marketable Features (MMFs) is a term often used alongside MMP. This refers to the process of releasing batches of functionality to customers, rather than pushing out individual features when they are ready. Again, the chunk of work is only considered ready for their customers once all of the stated features for that batch have been completed.

Who has financial responsibility for the project?

In the eyes of the team, the Product Owner has financial responsibility for the project. They are there to determine how the most value can be derived, and to relay that to the team in the form of a prioritised product backlog. In most cases, the Product Owner will undertake customer and market analysis to get the high-level picture, and will work with the team to break this down into deliverable slices.

Often the Product Owner is financially responsible by proxy. They will be acting on behalf of a budget holder from elsewhere in the business. This person may be referred to as the project sponsor, and will be the key stakeholder for outward communications, showcases and any project reporting. In such cases, the Product Owner must still have the authority to make decisions regarding the project or product.

Endnote

Although the Scrum framework is well defined, it is flexible so that it can be tuned to your individual circumstances. This is part of Scrum's beauty. Unfortunately, although Scrum seems simple in theory, putting it into practice is not so easy.

Between the three authors, we have been lucky enough to work in a variety of amazing environments, from the global fashion company where we first met to the UK Government, finance, travel, automotive, wine and many others. No two engagements are the same, yet time after time we are reminded how Agile and Scrum benefit the teams that we work with, and how hard teams strive to improve. In addition, while Scrum used to be the sole domain of software developers, this is no longer the case; it can be found working wonders in hospitals, publishing, finance and many other sectors.

So it saddens us when we hear people say things such as "Scrum doesn't work" or "Agile processes are a load of rubbish" because we know that isn't true. Scrum has been proven to work by many teams in many different environments. Unfortunately, there are a lot of bad implementations of Scrum, where people have lost sight of the Principles of Agile, or have assumed Scrum is as prescriptive as traditional methods. Agile and Scrum encourage you to think about how you work, not just about what you are working on. Being Agile can make a huge difference to organisations, departments, teams and individuals.

Our plea to you is to keep on trying. If you've had a bad experience in the past, take what you've learned from that and give Scrum another go. Continue to inspect and adapt, and focus your implementation on the values of Scrum and the principles of Agile.

They work. They can work for you too.

Acknowledgements

This book is the result of nearly three years' teamwork. Not just between David, Jim and Jit, but with input from many others.

Firstly, we would like to thank our families. David's wife and children put up with his crazy ideas and obsessiveness on a daily basis. Jim's girlfriend agreed to marry him during the writing of the book - we assume on the basis that it would be finished before they tied the knot. Jiten not only got married, but also became a father - the latter was considered a valid reason by the others for at least a few days off writing.

Next, we really appreciate the help we had from some early proofreaders who not only pointed out our grammatical errors, but also highlighted what worked and what didn't. These were Alex Vickers, Christina Ohanian, Malie Lalor and Sarah Henderson. Thanks also go to Paul Lowe for his professional proofreading of the final version of the book before we went to print.

The book was made to look beautiful by the highly talented Tom Mann. He put up with a lot of painful questions from the three authors and never complained when they asked for one more change.

Lastly, we would also like to thank all the people who have attended our talks and courses over the years (who asked the questions on which this book is based) and the individuals and organisations that we have worked with over the years (who gave us the experience we needed to answer the questions).

Bibliography

Beck, K. (1999). *Kent Beck's guide to better Smalltalk*. Cambridge,U.K.: Cambridge University Press.

Beck, K., Beedle, M., van Bennekum, A., Cockburn, A., Cunningham, W., Fowler, M., Grenning, J., Highsmith, J., Hunt, A., Jeffries, R., Kern, J., Marick, B., Martin, R., Mellor, S., Schwaber, K., Sutherland, J. and Thomas, D. (2001). *Manifesto for Agile Software Development*. [online] Agilemanifesto.org. Available at: http://agilemanifesto.org/ [Accessed 17 Dec. 2015].

Beck, K. (1999). *Extreme Programming Explained: Embrace Change*. Reading, MA: Addison-Wesley.

Cockburn, A. (2008). *Alistair.Cockburn.us | Elephant carpaccio*. [online] Available at: http://alistair.cockburn.us/Elephant+carpaccio [Accessed 19 Jul. 2016].

Cohn, M. (2006). *Agile estimating and planning*. Upper Saddle River, NJ: Prentice Hall Professional Technical Reference.

Greenleaf, R. (1970). *The Servant as Leader*. [ebook] Available at: http://www.benning.army.mil/infantry/199th/ocs/content/pdf/The%20Servant%20as%20Leader.pdf [Accessed 19 Jul. 2016].

Grenning, J. (2002). *Planning Poker or How to avoid analysis paralysis while release planning*. [PDF] Available at: https://www.renaissancesoftware.net/files/articles/PlanningPoker-v1.1.pdf [Accessed 17 Dec. 2015].

Kerth, N. (2001). *Project retrospectives: A Handbook for Team Reviews*. New York: Dorset House.

Kniberg, H. (2015). *Crisp's Blog » Elephant Carpaccio facilitation guide*. [online] Blog.crisp.se. Available at: http://blog.crisp.se/2013/07/25/henrikkniberg/elephant-carpaccio-facilitation-guide [Accessed 17 Dec. 2015].

Less.works. (2016). *Introduction to LeSS- Large Scale Scrum (LeSS)*. [online] Available at: http://less.works/less/framework/introduction.html [Accessed 21 Jul. 2016].

Miller, G. (1956). The magical number seven, plus or minus two:some limits on our capacity for processing information. *Psychological Review*, 63(2), pp.81-97.

Nationalcareersservice.direct.gov.uk. (n.d.). *Business project manager | Job profiles | National Careers Service*. [online] Available at:https://nationalcareersservice.direct.gov.uk/advice/planning/jobprofiles/Pages/projectmanager.aspx [Accessed 19 Jul. 2016].

Nonaka, I., Takeuchi, H. and Takeuchi, H. (1995). *The knowledge-creating company*. New York: Oxford University Press.

Oxford Dictionaries | English. (2016a).*ceremony | Definition of ceremony in English by Oxford Dictionaries* [online] Oxford University Press. Available at: http://www.oxforddictionaries.com/definition/english/ceremony [Accessed 18 Jul. 2016].

Oxford Dictionaries | English. (2016b). *ready | Definition of ready in English by Oxford Dictionaries*. [online] Oxford University Press. Available at: http://www.oxforddictionaries.com/definition/english/ready [Accessed 19 Jul. 2016].

Ries, E. (2011). *The lean startup: How Constant Innovation Creates Radically Successful Businesses*. London: Penguin Group.

Rising, L. (2008). *Questioning the Retrospective Prime Directive*. [Blog] InfoQ. Available at: https://www.infoq.com/articles/retrospective-prime-directive [Accessed 14 Jul. 2016].

Scaledagileframework.com. (2016). *About – Scaled Agile Framework*. [online] Available at: http://www.scaledagileframework.com/about/ [Accessed 20 Jul. 2016].

Schwaber, K. (2004). *Agile project management with Scrum*. Redmond, Wash.: Microsoft Press.

Schwaber, K. (1995). *SCRUM Development Process. In: 10th Annual Conference on Object-Oriented Programming Systems, Languages, and Applications*. [online] Austin, Texas, USA. Available at: http://www.jeffsutherland.org/oopsla/schwapub.pdf [Accessed 18 Jul. 2016]

Schwaber, K. and Sutherland, J. (2011). *The Scrum Guide: The Definitive Guide to Scrum: The Rules of the Game. Version 2*. [ebook] Available at: https://www.mitchlacey.com/resources/official-scrum-guide-current-and-past-versions [Accessed 19 Jul. 2016].

Schwaber, K. and Sutherland, J. (2016a). *The Scrum Guide: The Definitive Guide to Scrum: The Rules of the Game. 1st ed*. [ebook] Available at: http://www.scrumguides.org/docs/scrumguide/v2016/2016-Scrum-Guide-US.pdf [Accessed 18 Jul. 2016].

Schwaber, K. and Sutherland, J. (2016b). *Scrum Guide Refresh*. [video] Available at: https://www.scruminc.com/scrum-guide-refresh/ [Accessed 15 Jul. 2016].

Scrumalliance.org (2016). *Home - Scrum Alliance* [online] Available at: https://www.scrumalliance.org/ [Accessed 19 Jul. 2016].

Scrum.org. (2016). *The NexusTM Guide*. [online] Available at: https://www.scrum.org/Resources/The-Nexus-Guide [Accessed 21 Jul.2016].

Syncdev.com. (n.d.). *Minimum Viable Product | SyncDev*. [online] Available at: http://www.syncdev.com/minimum-viable-product/ [Accessed 19 Jul. 2016].

Takeuchi, H. and Nonaka, I. (1986). The New New Product Development Game. *Harvard Business Review*, January. Available at: https://hbr.org/1986/01/the-new-new-product-development-game [Accessed 11 Nov. 2016].

Wake, B. (2003). *INVEST in Good Stories, and SMART Tasks*.[online] XP123. Available at: http://xp123.com/articles/invest-in-good-stories-and-smart-tasks/ [Accessed 17 Dec. 2015].

Weaver, P. (2007). The Origins of Modern Project Management. In: *Fourth Annual PMI College of Scheduling Conference*. [online] Vancouver, Canada. Available at: https://mosaicprojects.com.au/PDF_Papers/P050_Origins_of_Modern_PM.pdf [Accessed 19 Jul. 2016].

Wells, D. (2015). *Extreme Programming: A Gentle Introduction*. [online] Extremeprogramming.org. Available at: http://www.extremeprogramming.org/ [Accessed 17 Dec. 2015].

Wikipedia.org (2009). *INVEST (mnemonic)*. [online] Available at: https://en.wikipedia.org/wiki/INVEST_(mnemonic)[Accessed 22 Sep.2017].

Wikipedia.org (n.d.). *Product manager*. [online] Available at: https://en.wikipedia.org/wiki/Product_manager [Accessed 19 Jul. 2016].

37960470R00076

Printed in Poland
by Amazon Fulfillment
Poland Sp. z o.o., Wrocław